CONCRETE

CONCRETE

FROM ANCIENT ORIGINS TO A PROBLEMATIC FUTURE

MARY SODERSTROM

University of Regina Press

Cover art: Modern concrete background by Bernulius/ Adobe Stock.

COVER AND TEXT DESIGN: Duncan Campbell, University of Regina Press
COPY EDITOR: Marionne Cronin
PROOFREADER: Kendra Ward
INDEXER: Sergey Lobachev, Brookfield Indexing Services

Library and Archives Canada Cataloguing in Publication
TITLE: Concrete : from ancient origins to a problematic future / Mary Soderstrom.
NAMES: Soderstrom, Mary, 1942– author.
DESCRIPTION: Includes bibliographical references and index.
IDENTIFIERS: Canadiana (print) 20200245627 | Canadiana (ebook) 20200245686 |
    ISBN 9780889777804 (softcover) | ISBN 9780889777866 (hardcover) |
ISBN 9780889777828 (PDF) | ISBN 9780889777842 (HTML)
SUBJECTS: LCSH: Concrete—History.
CLASSIFICATION: LCC TA439 .S63 2020 | DDC 620.1/36—dc23

University of Regina Press, University of Regina
Regina, Saskatchewan, Canada, S4S 0A2
TEL: (306) 585-4758 FAX: (306) 585-4699
WEB: www.uofrpress.ca

U OF R PRESS

We acknowledge the support of the Canada Council for the Arts for our publishing program. We acknowledge the financial support of the Government of Canada. / Nous reconnaissons l'appui financier du gouvernement du Canada. This publication was made possible with support from Creative Saskatchewan's Book Publishing Production Grant Program.

Canada Council   Conseil des Arts   Canadä   creative
for the Arts   du Canada   SASKATCHEWAN

For absent friends . . .

CONTENTS

Cement means concrete; concrete means stone; and stone spells eternity, so far as our finite minds can comprehend. The development of methods to manufacture great quantities of synthetic stone has given our present civilization the durability it so much needed . . . . The rapid progress of science and engineering in our country has been made possible by our cement mills, which have given us the essential material on which to found our prosperity.

—FLOYD W. PARSONS, 1924
History of the Portland Cement Industry, 2

CHAPTER 1
SOLID AS ROCK

"Concrete?" my friend said. "You're going to write a book about what?"

I suppose the reaction shouldn't have surprised me. Concrete has a bad rap in many quarters, particularly among people who pride themselves on being eco-friendly, nature loving, and sensitive to aesthetics. Many of my friends count themselves among them, and I suppose that most of the time I'd include myself on their team. Certainly my little inner-city garden with its billows of native plants in August suggests that.

But concrete is responsible, for good or ill, for the world we live in today, and not to appreciate it is to ignore both its menace and its beauty. Without it there would be no high-rises, no grand irrigation projects, no lettuce from California in the winter, no green golf courses in Arizona, a penury of hydroelectricity, more mud in some places, more solitude in others, and, very likely, little prospect of the disastrous climate change that is going to rain on our parade.

Don't believe me? Look out the nearest window, then try to imagine what the view would look like without concrete. Say your view is a cityscape; you'll have to mentally erase the apartment building across the street because the foundation is certainly made from concrete, and so might be the walls and the flooring. There'd be no sidewalks and perhaps no paved street. Underneath there'd likely be no sewer pipes, and unless the water pipes have been recently laid, none of them either. In fact, there might be no water coming your way at all, since in many places water comes from concrete storage dams that store water from rain and rivers.

Or perhaps you're in a suburb or the country. Even here the view would be drastically changed without concrete since low-rise, single-family houses are built on concrete foundations, even if their walls and roofs are made from other materials.

Then suppose the day is ending and twilight is falling on this trans-formed landscape. Time to switch on the lights—but no, in many places there would be no electricity because it is generated by water impounded in concrete dams. Nor would coal- or gas-powered elec-tricity generating plants be operating because their smokestacks are made of concrete. Nuclear power would also be non-existent since the guts of reactors have to be shielded by tons and tons of concrete.

So, in the semi-darkness you might decide to grab a snack to boost your spirits while you contemplate this fundamentally different world. Be prepared: the food available would be much less varied than what you're used to. There'd be nothing from irrigated fields, nothing that had been transported long distances by truck, nothing that had been stored and transformed in concrete warehouses and plants.

And don't even think of escaping this gloomy state of affairs by catching a sporting event or losing yourself in a monster show by a pop star: the stadiums and auditoriums they are held in are all made of concrete.

Convinced? Well, perhaps about concrete's usefulness. If you want more about its beauty—and its menace—just wait a bit.

The wonders of concrete—and I don't think the term is an exag-geration—became apparent to me during several years of researching and writing about urban spaces and urban life that culminated in my book *Road Through Time: The Story of Humanity on the Move* (Uni-versity of Regina Press, 2017).[1] Roads are about the most enduring creations we build, and where they go, goes change. Today, new paved roads are usually in part constructed of concrete, and as I travelled them, I began to think about what would happen if there were no concrete to build them. Then, as one thing leads to another, I began wondering what would happen if there were no concrete at all. That led me down a number of paths, to libraries, construction sites, dams, housing projects, cement plants, and modern monuments. At the end, I encountered the wall that our civilization appears to be headed for—disastrous climate change—and that concrete has helped construct as certainly as if it had been poured from jumbo concrete pumper trucks.

The first thing I realized was that concrete has its roots in deep time; in many respects it really is a "Rock of Ages." Cement, the material that holds concrete together, is in large part made from rocks laid down hundreds of millions of years ago when the shells and carapaces of organisms settled in the bottom of seas. Concrete can also last for thousands of years. Some Roman concrete is in great shape today: the dome of the Pantheon built in Rome circa 125 CE is still the largest concrete dome that isn't reinforced with steel. Just as impressive are the remains of a vast harbour installation constructed about 150 years before that off the coast of what is now Israel. While the docks and breakwaters are submerged today, the concrete is still strong, unlike that of many piers and wharfs constructed merely decades ago.

The second thing I learned as I tried to tease out both the history and chemistry of concrete is that what is considered concrete has changed over the centuries. Today, concrete is defined as the substance created by a judicious mix of a binder (usually a variation of Portland cement), sand and/or aggregate, perhaps a couple of additives, and water. The slurry produced can be poured into forms, where it hardens. Through a chemical reaction, the cementitious material binds the aggregate together by making a lattice-like web of molecules that hardens over time, sort of the way water forms ice crystals as it freezes. In concrete, after twenty-four hours or so, the transformed mixture will be hard enough in most instances that forms can be removed. However, it won't usually reach its design strength for twenty-eight days and may continue to gain strength well after as it ages.

Today, we don't know how long it took for the concrete the Romans made to set, and we are only now unravelling some of its chemical

Figure I.I: Rome's Pantheon, built nearly 1,900 years ago, is still the largest unreinforced dome in the world. Painting by Giovanni Paolo Panini, ca. 1730. Public Domain.

Figure I.2: Roman walls, like this one at Conímbriga in Portugal, were made with a concrete core and stone facing. Photo: Mary Soderstrom.

mysteries. Sometimes they poured it, as when they made harbour installations, but most often it was made and used quite differently from the way it's used today. Basically, it was mortar made from fired limestone, water, and ordinary sand that was frequently mixed with volcanic sand or ground-up bricks. It was then slathered onto or poured over rocks of various sizes; our word *cement*, which now refers to the substance that makes concrete stick together, comes from the Latin word for what we'd call rubble or aggregate: *caementum*.

But the Romans were far from the first people to discover one of the key elements in their magic mix, because people all over the world again and again stumbled on the amazing properties of fired limestone, also called quicklime. Each time must have been literally an explosive surprise because a sudden, intense reaction that sets the water and the rock raging is what happens when you add superheated limestone to water. Some historians suggest that a lightning strike that burned limestone set the scene for the discovery. Or perhaps the process was encountered when early humans heated rocks to make it easier to work flint or obsidian in toolmaking: there's evidence that fire was used to crack rocks as far back as 20,000 BCE.[2]

Or perhaps—and this seems more likely to me, given it's clear that lime mortar was made well before techniques for making pottery were developed—the discovery is related to the way many Indigenous Peoples have cooked food. You may have done it yourself on a Scout excursion or at summer camp. You begin by filling a container—a pouch made of hide, perhaps, or a basket of tightly woven grasses, or a vessel made of bark stripped from a tree—with water, grains, and perhaps meat. Then you heat stones on a fire until they are very hot. When you drop them into the container, what happens ordinarily is that the water heats and after a while the food cooks to become porridge or stew. But if certain kinds of rock were forgotten in the fire and grew hot enough, a critical stage would be reached. Voilà, instead of supper you had an explosion of boiling water.

In the terms of modern chemistry, this is what happens: when limestone or other material, such as shells, rich in calcium carbonate (a compound that has one calcium atom, one carbon atom, and three oxygen atoms) is heated to above 850 degrees c, the bonds holding molecules together are loosened. Molecules of carbon dioxide (CO_2) are released, while the remaining molecules of calcium and oxygen form calcium oxide (CaO, or quicklime). The rocks or shells are considerably lighter than they were before firing. Put them in water, though, and the unstable quicklime reacts with the hydrogen and oxygen in water to produce a new substance, calcium hydroxide $Ca(OH)_2$, and a lot of heat energy.

Doing that now is pretty impressive, but imagine what it must have been like in a time when people had no receptacle that would withstand the heat of fire long enough for water to boil on its own. (This might also explain how several cultures learned to use quicklime to process food. Among them are several North American Indigenous Peoples who used it to prepare corn for eating and storing, and to whom we owe both hominy and *masa harina*, the corn flour used in tortillas.)

You could do this yourself this weekend, if you dared. There's one recent YouTube video which gives you step-by-step instructions. A cool dude with piercings, dreads, and half-shaved scalp shows how to make a kiln from straw, muddy clay, and water. "It's fun and cool," he repeats again and again before he gets to work. The wet, clay-soaked straw he forms into a metre-tall chimney. Then he puts in kindling and sets fire to it so that the clay is baked. That done, he piles in layers of limestone rubble and oyster shells that he's scored off a friend who runs a restaurant, alternating with wood. Then he lights it, and after several hours, during which he tops up the fuel occasionally, the shells and rubble have been reduced to a fraction of their weight and can be easily crumbled.

The fun part follows: he pours water over the residue, and the mixture begins to boil ferociously.[3] The process is fascinating but scary. Other videos about making quicklime feature scientists wearing special goggles and protective gear, who warn that this is something not to be attempted at home. Not this guy, even though the stuff is dangerous: the life-threatening potential of the substance was well known to the Nazis, who at least once used it to kill people who were thrown alive into a pit containing it.[4]

But early—and less bloodthirsty—experimenters also found that when the chemical reaction had stopped, the result—called slaked lime—became a kind of paste which, when painted on walls, left a pleasing layer that could be drawn upon, and which offered some protection from the rain and snow. (The nineteenth-century North American version was called whitewash, and anyone who's read *Tom Sawyer* will be familiar with it, since Tom masterfully schemed to get his friends to take over whitewashing a fence.)

Excavations of dozens of settlements in the Middle East have turned up remnants of lime plaster dating as far back as 8700 to 7000 BCE, to the time when humans were just beginning to build towns.[5] One of these settlements is Asilkli Höyük, in the central Turkish region of Anatolia. Houses there were frequently made of mud brick, but limestone blocks held together with lime mortar were also used for wall construction. In what is now Jordan and Israel, two other settlements that date from around the same period have hundreds of square metres of floor plastered in a hard lime-based substance. Burning that much limestone to make quicklime must have required a great amount of wood, since it takes about 3.5 to 8 tons of wood to produce enough quicklime for one house. Obviously, the energy requirements for making building materials then had an effect on the environment, just as those for making cement do today, as we shall see.

People elsewhere in the world also discovered the wondrous transformation wrought by burning certain kinds of rock. Examples include ancient buildings in Gansu Province in north-central China that date from about 5000 BCE. When excavated, the residential complex had floors that were as strong as modern concrete but were made of fired ginger nut (a type of stone composed of calcium carbonate and clay) with *kunkur*, a coarse limestone, as an aggregate.[6] Chinese builders also soon experimented with adding other materials to the mix: one of the most successful was sticky rice, a short-grain rice that becomes gummy when it is cooked. Records show that it has been used since at least 1,500 years ago to make a sort of concrete that gets stronger as it ages.[7]

On the Indian subcontinent, lime plaster was used as far back as 3000 BCE in Mohenjo-Daro, the ancient city of the Indus Valley where extensive drainage and water delivery systems were developed.[8] Around the same time, the Egyptians were building the first pyramid complex at Giza from limestone blocks, and while they mostly used gypsum for plaster, which doesn't require the same high temperatures to make, in some places lime plaster was used.[9]

Mixed with sand to make lime mortar, the slaked lime slurry securely held big rocks in place to make walls. In the Western Hemisphere, the Hohokam, a North American Indigenous People who lived in what is now Arizona between 1000 and 1500 CE, used a concrete-like mixture of natural calcium carbonate, clay, and sand to build their large houses.[10] In Central America, the Maya in some places also used a cement-like mortar.[11]

Another use of lime in vernacular architecture is in what is called *tabby* in North America. Its origins may be in Africa. Possibly the Spanish encountered it there—the name appears to come from the Spanish word *tapia*, which refers to earth compacted between boards— and brought it with them during their explorations. Used extensively in the southern United States and the Caribbean, it was made from burned and ground oyster shells that have as much or more calcium carbonate as limestone. Mixed with sand and water, tabby could be poured like modern concrete to make columns and bricks.[12]

But the cement or mortar made only with quicklime has a couple of big drawbacks: it takes forever to dry and it can't be used underwater or where it is damp. The Greeks discovered that they could get around this if the sand they used to make the mortar was of volcanic origin, like that from the island of Santorini, which had been rocked by enormous volcanic eruptions. The resulting material hardened in the presence of water and dried much more quickly, although the Greeks had no idea why. The Romans, great pragmatists that they were, did not know either, but they found that the ash from Pozzuoli near Mount Vesuvius, particularly those particles that were glassy and very fine, produced the same effect.

The secrets of the chemical reactions would have to wait centuries before being discovered. The Romans' approach was much like that of nineteenth-century tinkerers and engineers who had little idea of the chemistry and physics behind the "artificial stone" (a frequently used term) they were trying to make. In 1759, when English engineer John Smeaton was building the Eddystone Lighthouse, the first grand modern project that used a cement mortar that hardened under water, Enlightenment scientists were just beginning to identify chemical elements as we know them today.[13] What lay behind the particular properties of each element and the reasons why they react with each other the way they do wouldn't be understood until 150 years later, when the atomic structure of each chemical element had been decoded. We'll discuss later how changing the technique of making lime mortar by adding a few other components makes modern cement and modern concrete. Suffice for the moment to say that today the amount of modern cement and concrete produced each year is phenomenal. In many ways it also is a truly egalitarian material, used by multinational companies and all layers of government to build everything from skyscrapers to hydroelectricity projects, as well as by poor squatters to build homes that others might call hovels.

How to get a handle on this vast subject, though? Perhaps by going back to the beginning to look at cement and the concrete that it makes through the lens of the Roman's world view, with its foundation in the four elements that they thought made up the universe: earth, fire, water, and air.

All four of the Roman elements were vividly on display in the late-May morning in 2017 when I began this investigation in earnest with a visit to the little town of Port Daniel-Gascons on Quebec's Gaspé coast and a tour of the newly completed McInnis Cement plant. It is the first new cement plant to open in North America in decades, and despite its location thousands of kilometres and millennia of time away from the Roman origins of concrete, it seemed to me that a closer look might say a lot about both the history of concrete and its future.[14] Just inside the new cement-making complex's boundary stood the mountain of *earth* that had been removed to reach the limestone underneath, the bedrock of the rugged highland. *Air*, remarkable in its clarity after several days of rain, transmitted the colours of early spring brilliantly. *Water* stretched from the bottom of the hill as far as the eye could see: this part of the Gaspé Peninsula marks the beginning of deep water in the Gulf of St. Lawrence. And *fire* was being offloaded at the dock in the form of petroleum coke to fire the great furnaces that would burn the limestone to make the cement.[15]

Likewise this journey of discovery will pass through four sections in which important aspects of the concrete story will be considered as they relate to the four elements. Chapter 2, "Earth," will deal with the materials that go into making cement and concrete, and the history of how they were discovered. Chapter 3, "Fire," contains stories both menacing and comforting: how concrete is born in fire, is an essential element for warfare, and, paradoxically, builds the places where we light our home fires. Chapter 4, "Water," talks about the multiple ways concrete allows water to generate electricity, irrigate fields, and sustain life. Chapter 5, "Air," is divided into two parts. One details concrete's menace through direct and indirect CO_2 emissions. The other considers the way concrete allows the human spirit to take wing.

The concluding chapter, entitled "The End of the Road," brings us back to McInnis Cement and reflections on what paths we must take now to control the menace and promise of concrete.

So let's hit the road (a concrete one, of course) . . .

CHAPTER 2
EARTH

IT'S ELEMENTARY

Of course, you might say, to connect today's concrete with the Roman's element earth is a no-brainer. It's dirty and what it's made of comes from the earth. But I wasn't thinking of that when I pulled into the parking lot at McInnis Cement on the Gaspé coast of Quebec.

Let's be clear: the plant is located where it is for two earthy reasons. First, the hill behind it contains limestone of very good quality, and in quantities large enough to last fifty years or so.[1] Second, local politicians wanted the project to win votes. Some would say that it was a "dirty" project from the beginning, in many senses of the term.[2]

Maryse Tremblay met me in her four-by-four in the employee parking lot, across from the great pile of overburden that had been removed to get to the limestone. The pile was covered with a light sprinkling of very green grass, sown to stabilize the hillside but not yet hiding the tons and tons of smashed rock stored there. Tremblay, whose title is director of communications and corporate social responsibility, explained that environmental protection regulations prohibited

trucking the earth away—but then where could it be trucked anyway? Nobody would want a load of this waste in their backyard.

The Gaspé Peninsula, on whose south shore the plant is located, is a region of extraordinary beauty where Indigenous people lived relatively lightly on the land for several thousand years before Europeans arrived. The French explorer Jacques Cartier made landfall here in 1534, coming ashore on July 12 at Port Daniel, where one of the first Masses celebrated in North America was held.[3]

European settlement did not follow immediately. Cartier had been mistaken to think that the waters south of the peninsula were actually the opening of the St. Lawrence and hence the way to the interior of the continent, or even, he hoped, a passage to the Orient. When Cartier reported his error back home, the French decided they were not very interested in the region.

On his return trip a year later, Cartier went farther north and found the true entrance to the St. Lawrence, which King Louis XIII and his key advisor Cardinal Richelieu found much more intriguing. Nevertheless, some French fishermen did settle on the Gaspé coast, where the warm waters of the Baie de Chaleurs teemed with fish. But in the late eighteenth century, when the British gained control of New France,

Figure 2.1: The brand-new McInnis Cement plant in May 2017. Photo: Mary Soderstrom.

they left. In the following decades their place was taken by sojourners from other seaports, including the island of Jersey in the English Channel. At the same time, the first farming communities were established. The importance of the rock that formed the backbone of the peninsula was recognized rather quickly after that: the limestone was of such good quality that, as early as 1836, boats from Prince Edward Island were putting into little Port Daniel to take on lime made from the rock, as well as to cut blocks of limestone for building.[4]

The ordinary folk of the Gaspé Peninsula were never prosperous, although a few fortunes were made from exploiting fishing, mines, and forest resources. At the turn of the twentieth century a railroad line running along the southern coast was finished, connecting the town of Gaspé at the eastern tip of the peninsula with the main line of the Canadian National Railway (CN). For nearly 100 years it carried products and people all along the coast, providing a relatively quick and sure link to the rest of Quebec and the greater world. As recently as 2013, three evenings a week you could catch the train at Port Daniel and be in Montreal by noon the next day.

But the train option is no longer available. Today, the 934-kilometre distance translates to a minimum of twelve hours of steady driving by car at the speed limit, and you'll arrive exhausted, not rested, after the drive. The official reason Via Rail gave for suspending train service was "rail infrastructure problems."[5] It's true that some of those problems were due to years of lack of maintenance by the rail company that owned the tracks. But please note: the closure came just at the moment when McInnis Cement was rapidly moving toward approval and construction of its new plant. Although the rail link was touted in the McInnis prospectus as one of the ways cement could get to market, part of the track that crossed the cement plant's territory was torn up early in the project, presumably to make construction easier.

The rail line will be rebuilt, Maryse Tremblay assured me as we stood looking west along what had been a stretch of track. In the meantime, cement must be trucked about 80 kilometres to New Richmond to meet a still-operating CN line. Nevertheless, the commitment to rail is real, she said.[6] Not long after production began, McInnis acquired twenty-five new rail cars designed to transport, load, and unload dry bulk material under tightly sealed conditions.[7] Then, shortly after my visit, the Quebec provincial government promised c$100 million toward the necessary repairs to the rail line, although rehabilitation work didn't start until the summer of 2019.[8] Some observers see this as just another example of politically motivated, but economically doubtful, support for a plant—and a region—that doesn't have much going for it now but beautiful scenery and—to put it baldly—dirt.

Locals had welcomed the news that the cement plant would be built. Unemployment was very high: in 2006, when McInnis was preparing its explanatory documents, the unemployment rate was 23.1 percent in the Municipality of Port Daniel-Gascons, more than three times the provincial rate of 7 percent. Median household income was also the lowest on the peninsula: c$30,719.[9] Industry in the region had been decimated during the previous several years as paper mills and cardboard manufacturers closed up shop.[10] The great softwood lumber forests of the interior had been chopped down too: the port of Chandler, forty-five minutes by winding road to the east of Port Daniel, once shipped newsprint manufactured from locally cut trees to

The New York Times. But that had passed, victim of both the immense changes in the information industries and the end of easily harvested forests. The collapse of the cod fishery twenty years before had been a hard blow too.

But the May days I was there were unusually hopeful.[11] The little ports along the coast with their seafood processing plants were humming with activity. It was a bountiful season for snow crab in these waters, and boats were putting out in great numbers to set traps, haul them in, and see the catch sent to market in the rest of Canada and the United States. The lobster season would follow, and thousands of dollars would change hands on the waterfront. The catch would not only mean money right now but also possible eligibility for unemployment benefits after the fishing season was over, valuable assistance when unemployment rates continued high.[12] Some of the people on the coast would soon go to work in the tourist industry also. With the continued low Canadian dollar and concern about erratic politics in the United States, it looked like the summer would be a banner year for travellers visiting the scenery and delights of Gaspesia.

There also was some money still floating around from construction of the McInnis plant. Heavy machine operators, engineers, and cement experts had crowded lodgings for more than three years while the plant was being built. The woman who ran the bed and breakfast where I stayed proudly showed off her new, top-of-the-line refrigerator with the freezer on the bottom, bought with the extra cash a constant stream of guests on generous per diems had brought in.

But only a handful of employees would be on-site once production was revved up. Three teams of six control-room workers—local men who'd been specially trained for their highly technical jobs, Tremblay said—the cleaning and maintenance staff, and a few others would be the regular work force.

Cleaning and maintenance staff. Another kind of dirty work, in other words.

But working with concrete is dirty, almost by definition. The raw material that makes roads and dams looks pretty dirty from the get-go. My first contact with cement—as I'm sure it was for a generation of people who grew up with parents trying to upgrade small houses—was with the sacks that my father would lug home from the hardware store in the trunk of the car. There were three kinds: cement sacks, sand sacks, and gravel sacks. All three were far too heavy for a kid like me to carry when my parents were in the middle of their do-it-yourself phase.

My dad would dump some of the cement—he used a number two size tomato can to measure it—into the trough that he'd acquired somewhere. Then he would add some sand and gravel and, finally, water. He'd mix it together with a long-handled hoe, periodically testing the consistency. Then he'd ladle it into the forms that he'd built to shape the steps or whatever he was making. A couple of times he allowed my sister and me to smooth the concrete with a trowel, and once he pressed our hands into the slab, after which he scratched our names and the date underneath. When he was finished, the little sand or gravel that remained might be dumped in a pile at the end of the yard, out by the alley, where we could play with it, making a town and building hills and playgrounds with bits of wood and scrap metal, racing the little cars that belonged to the neighbour boy. But we never were allowed to even touch the cement that might remain.

It looked like gray flour. It looked like it might feel smooth and fluffy to the touch. Why couldn't we mix it up with water, turn it into mud pies for consumption by the imaginary creatures who lived in the village we'd built? At the time I thought the restriction was just another example of adults not understanding what kids wanted to do.

The cement powder, while it might look a lot like any other kind of dirt, was different, though. It wasn't the natural, if unwanted, dirt that collected in corners of the house or that we were always tracking in from outside. It was, and is, carefully, expensively crafted from substances that aren't that uncommon, which is the reason why once its secrets were unlocked it could be made all around the world. No, it wasn't because cement was such a rare substance that we weren't allowed to play with what was left over. The reason was because what you got when you added water was a very special sort of "mud." It would not only be hard to remove once it dried, but it also became very hot almost immediately.

The first step toward understanding what that process meant didn't come to me until I was studying English literature as an undergraduate and read *As I Lay Dying* by William Faulkner. The novel seems a long way removed from building anything: it's the story of a poor, white family in Arkansas whose mother has died. They want to bury her with her people, and so they put her coffin on a wagon and start the several-day trek to her hometown. But the coffin slips and falls on the leg of one of the sons, breaking it. They splint the leg and load him in the wagon too, but as they jolt their way along the unpaved country roads that are little more than ruts, he is in great pain. Their solution: convince a storekeeper to sell them ten-cents worth of cement, which they mix with sand and water to pour on his leg, with the idea that when it sets it will hold the broken bones in place.

What they don't reckon on is the heat. At first the watery mixture is cool, but it quickly heats up.

> After supper Cash began to sweat a little. "It's getting a little hot," he said. "It was the sun shining on it all day, I reckon."
>
> "You want some water poured on it?" we say. "Maybe that will ease it some."
>
> "I'd be obliged," Cash said . . . .
>
> So we poured the water over it. His leg and foot below the cement looked like they had been boiled. "Does that feel better?" we said.
>
> "I'm obliged," Cash said. "It feels fine."[13]

But of course it isn't fine, and when they get to their destination and a country doctor removes the hardened concrete, most of the skin on Cash's leg goes too. He'll not be able to walk for months afterwards, if ever again. The doctor is incredulous: "Concrete . . . God Almighty, why didn't [your father] carry you to the nearest sawmill and stick your leg in the saw? That would have cured it. Then you all could have stuck his head in the saw and cured a whole family . . . ."[14]

After reading that, I put it out of my mind until my son broke his hand when he was about twelve in a rock-climbing accident that was both silly and a tribute to his daring. As we waited in the emergency department for the technician to prepare a plaster cast, Faulkner's scene

Figure 2.2: Storage pyramids at McInnis Cement.
Photo: Mary Soderstrom.

floated up from the depths of my memory. Needless to say, I was reassured when the tech explained that the cast he would make was basically a dressing impregnated with plaster of Paris, which is ground, kilned gypsum, not cement. When water was added, the mixture would heat up, but not nearly as much as a mixture of cement and water would. The cast would also soften in water, so my son was to be careful washing and bathing, something that a kid his age wouldn't be bothered by at all. And when it was cut off, he'd be none the worse for wear.

Thank goodness, I thought, as I realized just why my own father was so adamant about us not playing with leftover cement "dirt."

The McInnis plant was designed to keep that dirt under control. As Maryse Tremblay gave me the number one tour, a water truck rolled back and forth across the site, like a Zamboni making ice in a hockey rink. Keeping dust down was an important concern when planning the complex: it comes up a couple of dozen times in the detailed proposal produced just before construction began. Part of the concern is about clouds of pulverized rock that might be emitted when rock is blasted and then broken up at the face of the limestone deposit. But also, once the rock is crushed even smaller, no one wants the dust to escape—for health reasons and because at that point it is becoming a valuable commodity. Some buildings at the plant have lower air pressure inside than outside in order to keep dust inside. Once the cement has been finished, the air pressure difference is maintained as the cement is forced through large, sealed pipes to waiting ships or trucks.

When I was there, it must be noted, everything was shiny and new. The mammoth concrete silos and buildings that dwarfed the men and machinery working on the ground were still dazzlingly white. Two storage pyramids were Ikea blue and looked as if they'd been inspired by the I.M. Pei glass pyramid that now is the entrance to the Louvre Museum in Paris. Yellow ladders climbed up the side of the structures, showing no rust, no smudges, no indication they were anything other than vertical and diagonal elements in an architect's dream. How long this pristine state would last was anybody's guess . . .

ROMAN DIRT

But dirt, in the form of sand from a riverbank or quarry, lay behind the success of Roman concrete. Making mortar was such a tricky thing that it required a long apprenticeship—one early text says that a master mason could not take on an apprentice for less than six years because it took that long to learn the art. And art it was, far from a science. How to choose the right kind of rocks to be burnt, how long to fire them, when to allow the fire to go out: these were all things that a master taught his apprentices.

The breakthrough, the great technological change if you will, came when Roman builders started using sand from volcanic rocks systematically.[15] At what point this happened is unclear. As I said earlier, Greek builders also had used this kind of sand to make mortar for some projects, but even though the concrete produced was clearly superior, they seemed not to realize what caused the difference in the end product. They had no way of knowing that quicklime mortar and hydraulic cement harden through two completely different chemical processes.

In the first, lime mortar absorbs CO_2 from the air and the mix slowly becomes calcium carbonate, much like the rock that was burned to make the lime in the first place. In the second, the reaction takes place in the presence of water, and is considerably quicker to set. By the third century BCE the Roman artisans and engineers had learned by trial and error the right sand to use and the right proportions of it compared to other ingredients. (The exact chemical reaction that took place in the Roman's version is still being puzzled out by scientists.[16] In the summer of 2017, a group of researchers reported that unique chemical features in their raw materials meant that Roman concrete gains strength for decades, or even longer, as it ages.)

Mount Pozzuoli near Naples is where this sand came from, and pozzolana or pozzolans have become the names for all materials that contain forms of silica and alumina that are soluble in water and react with it to make a cement that sets in water. Some pozzolans are natural, including several other kinds of volcanic rocks. When building in their northern territories the Romans used ground Rhenish tuff found near the Rhine River, which is also of volcanic origin. There's evidence that the Mayans stumbled on similar volcanic sand when making lime mortars for their temples.[17] Romans found later that using ground-up terracotta or fired bricks also worked, but pozzolana forged in the fires of the earth worked best.

The Romans used concrete for major buildings and the dwellings of the wealthy, but it appears that before the Great Fire in 64 CE during Nero's reign, much construction was done in more flammable material. The fire burned for six days, wreaking widespread destruction, according to Tacitus.[18] Contrary to popular belief, though, Nero did not set fire to the city, nor did he fiddle while it burned. But he may have been pleased once the flames were quenched, because the burnt-over sections of the city lay ready for monument building. Most commentators date a major change in construction technique from the rebuilding, because afterwards concrete and fire-resistant materials were insisted upon.

The ruins of Pompeii—the city destroyed in the eruption of Mount Vesuvius only a few years later, in 79 CE—show that concrete was used in many constructions there, although more modest buildings built with cheaper, more flammable materials would have been completely destroyed by fire after the volcano's explosion.[19]

Vesuvius is only about fifty kilometres away from Mount Pozzuoli, whence came the volcanic sand that was the secret of Roman concrete, and it's possible that the builders of Pompeii might have used a more local variety. But pozzolans from Mount Pozzuoli were considered the very best for building big projects. To construct the harbour installations at Caesarea Maritima on the coast of what is now northern Israel, the Romans imported about 52,000 tons of pozzolans between

23 and 12 BCE.[20] That was in the days of Herod the Great, when the Middle East supplied great quantities of grain to Rome. Providing a safe harbour for this trade was a major reason for building the installations. Wood for forms to mould the concrete, as well as pozzolans, were carried to Judea at least 150 times as ballast in ships returning from the Italian peninsula after carrying grain there.[21]

An early work on architecture that mentioned using pozzolans as additives was written around this time by Vitruvius, a name that has become famous in the architectural and engineering world. His *Ten Books on Architecture* was lost for more than 1,000 years, until the work was rediscovered during the Renaissance. In 1515 Fra Giovanni Giacondo published the first Latin translation, which was widely circulated and which influenced many Renaissance architects and builders.[22]

Today, no document dating earlier than Giacondo's translation exists that refers to using pozzolans. It's obvious, however, that this knowledge lived on in some places after the fall of Rome. The Gothic period was a time of much castle and cathedral construction. Nineteenth-century archaeologists saw the stairways and arches in London's White Tower, built by the order of William the Conquer in 1078, and called the rubble-mortar mix that made them "concrete."[23] So strong was a similar substance laid between the stone walls of the Reading Abbey in 1130 that the "concrete" core remained long after the stone had stripped away.[24] Ninety years after the abbey was built, it's clear that artisans who began building Salisbury Cathedral in 1220 knew how to render lime mortar into something like hydraulic concrete.[25] The water table is only a few metres under the building, yet recent research shows that the lowest levels of foundation are made from a mixture of "solid" limestone mortar.[26] The builders also filled the walls—formed of two layers of carefully cut stone with a hollow between—with a rubble core consisting of pieces of chalk mixed with mortar, in the manner of many Roman constructions.[27] Similarly, in France, the Cathedral at Amiens, where construction also started in 1220, has a twenty-five-centimetre concrete-like slab sandwiched between layers of brick and other stone as part of its foundation.[28]

In the south of France it appears that practical knowledge of how to make waterproof concrete was also passed from craftsman to craftsman (or craftswoman to craftswoman, according to historian Chandra Mukerji) because it was useful in keeping the many Roman accomplishments like aqueducts and canals in good repair.[29] When the utility of adding pozzolans to cement to make hydraulic concrete was rediscovered in the Renaissance, engineers building the Canal du Midi, which crosses France from the Mediterranean to the Bay of Biscay on the Atlantic, found workers who already knew how to work with it.[30]

But, in general, the idea of adding special kinds of "dirt" to the mix of sand and lime mortar to make it set faster and resist water seems to have disappeared for centuries. One of the reasons may have been the scarcity of materials with properties like those of the natural pozzolans of Italy.[31] This was true even where the ability to set in the presence of water would have been a great advantage, as in the Netherlands. Although serious attempts to reclaim lowlands from the sea began in the late Middle Ages, it wasn't until the mid-eighteenth century that the Dutch began experiments using *trass*, their name for the Rhenish tuff that had been used for building by the Romans for several hundred years at the beginning of the modern era.[32] The impetus for the Dutch experimentation was an infestation by naval shipworm, a mollusc that attacked all wooden structures, including dikes, which were then generally made of stacked peat and wood. The same scourge affected docks and breakwaters in Venice, where engineers turned to pozzolans to enrich lime mortars, making a slurry of it with small stones and pouring it into forms or "caissons."[33] Italian architectural historian Robert Gargiani asserts that this experience prepared the way for understanding how to use modern concrete because a key element in concrete work now is the idea of pouring the slurry into forms. He suggests that this was part of a century-long process of relearning the secrets of the Romans, which ultimately led to understanding that some dirt, sand, or ground rock was better than others in making what came to be known as hydraulic cement; why that was the case remained a mystery for a couple of centuries longer.

Knowledge about what the Romans had done and how that could be used by contemporary builders began to circulate throughout Europe in the eighteenth century. The English engineer John Smeaton was particularly inspired by works by French engineers, including Bernard Forest de Bélidor, who wrote a treatise on architecture that picked up where earlier translators of Vitruvius had left off concerning the efficacy of pozzolans. Smeaton also visited Holland, where trass was then being used frequently, and began systematically to study various ways of making mortar. This work was essential when, in the mid-1750s, he received the contract to build a lighthouse at Devon off the south coast of England, the Eddystone Lighthouse.

The Eddystone rocks were a major peril to navigation, and two attempts had been made previously to build a lighthouse. Both were wooden structures, built with the idea that they would move harmlessly with the winds and tide. But both had failed, and Smeaton opted for solidity instead. It was a tough job, particularly since work could proceed only during the summer and at low tide.[34]

To get quick-setting mortar that worked in water, he tried mixing quicklime with trass, but found the result unsatisfactory. (His journals, by the way, read like those of a dedicated, scientific mud-pie maker: he essentially mixed up small quantities of various kinds of materials and then recorded what happened under various circumstances.) After Smeaton's experiments with trass, he discovered a shipment of pozzolans, which he'd read about in French writers, that was available at a "good price." The material had apparently been imported to Britain for repairs on a bridge over the Thames, but hadn't been used. His experiments with them showed they worked much better than trass. The mortar he concocted with them was so effective that the mythic lighthouse, completed in 1759, served for 127 years before being decommissioned. Thirty years after completion, Smeaton noted, the mortar underwater was holding well but that the rock above the high-water line was "sensibly corroded."[35] In the end, it was wave action undercutting the very rock on which the lighthouse stood that doomed it, but not before its steadfastness became emblematic.[36]

Elsewhere during the same period, tinkerers, and some engineers, discovered that certain naturally occurring sorts of rock had pozzolan-like properties. They were used to make a "natural cement" fired at temperatures of 800 to 1,100 degrees C, temperatures that are not much higher than those necessary to make lime from limestone or

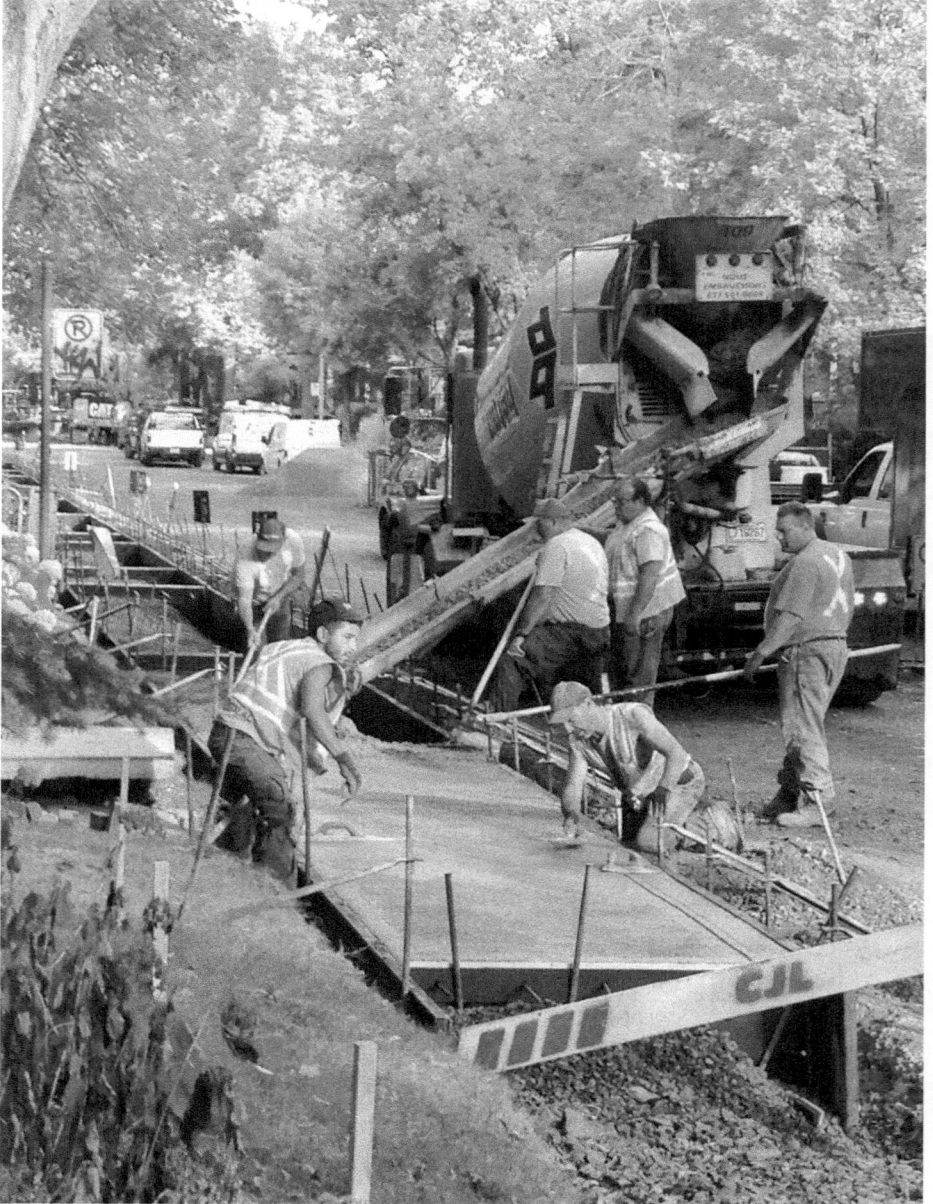

Figure 2.3: Concrete must be placed within ninety minutes of adding water to the cement and aggregate mix. Photo: Mary Soderstrom.

shells. In Britain this was called "Roman cement" and used marl—a kind of limestone that contains clay—as the source of the lime.[37] The proportions were something like 10 percent pure lime made from the marl, and around 25 percent clay containing silica and alumina. Some of its formulations set extremely quickly, as the directions on one container explained: "To use this CEMENT, mix it up with water to the Consistence of a thick Paste, and apply it immediately . . . . Wet no more at a Time than can be used in 10 or 15 Minutes. After it has once set, it will not be fit for Use again . . . ."

Natural cement was widely used through the middle of the nineteenth century in a range of projects as varied as tunnels under rivers and ornamental statuary. In North America, a source of the right kind of stone was found early in the nineteenth century in New York State. It was used in some of the era's biggest engineering projects: canal networks that allowed the transport of goods and material from place to place faster and more easily than by cart. Eclipsed by railroads within a few decades, inland waterways nevertheless became as mythic as the Eddystone Lighthouse, with songs about the Erie Canal entering into the bedrock of American culture.

But determining just why some substances made things stick together, and others didn't, remained shrouded in mystery. Among the artisans and engineers who undertook systematic research into the problem was the Frenchman Louis Vicat, a young military engineer who studied what was being done in the Netherlands and Italy. He began to suspect that the key to hydraulic cement lay in compounds that were created in the heat of volcanic action. Over nearly a decade—while he waited to get financing for a bridge project whose construction he was supposed to supervise—he experimented with different processes through which silica and alumina are combined with lime at great heat.[38] The materials had to be mixed together, dried, and then burned to get a substance that set in water: firing the materials separately didn't work. The trick appeared to be in mimicking the inferno of volcanic action, Vicat found. He developed kilns for first burning—calcining is the term frequently used—the lime and other ingredients together, then grinding the clinker produced. The temperatures required for doing this are nearly twice as high as that needed to make quicklime, about 1,450 degrees C, and therein lie two of the great problems created by our enormous use of modern concrete: where to get the energy to attain those temperatures, and what to do with the greenhouse gases emitted in the process.

At the beginning of the nineteenth century, those concerns were not considered of great importance. What was of greater concern was finding a way to make concrete consistently as good or better than the Romans. Louis Vicat didn't patent his invention but donated it to the world because he felt so strongly about the value of the compounds he produced.[39] For this act, he received much credit in France: in 1839 Balzac wrote in his novel *Le curé du village*, "What shall Vicat's recompense be . . . he who has alone achieved real progress in the practical science of construction?"[40] Elsewhere, though, the lack of patent, whatever Vicat's noble motives, has obscured his contribution.

French and British tinkerers were not the only men intent on uncovering the secret of concrete, by the way. At the same time as Vicat was doing his experiments, the Georgian engineer Egor Gerasimovich Chelidze (also known by his Russian name, Egor Cheliev) created a new binder by mixing lime and clay. His results were published in 1825 in *Complete Instructions on the Preparation of a Cheap and Superior Mortar or Cement of Great Strength for Underwater Constructions.*[41]

Two years later, the first British patent for cement as we know it was awarded to Joseph Aspdin, who called his binder "Portland cement" because the material was designed to substitute for the renowned building stone from Portland, another island off the English coast.[42]

But it was much later in the nineteenth century before concrete began to be widely used, even though the raw ingredients can be found all over the world.

The need for cement and concrete increased as good, easily accessible sources of building stone became exhausted and the forests of eastern North America were cut down to house the burgeoning population. Seeing immense opportunities, dreamers and inventers took up the challenge: in the mid- to late nineteenth century there were hundreds of patents issued for cement and concrete manufacture. The American Thomas Edison was one of those fascinated by the material and was eager to find the best methods of producing it. Famed for his invention of the successful incandescent light bulb, he also received forty-nine patents for making cement and using it.[43]

For decades after the secrets of making Portland cement were unravelled, much cement was produced close to where it would be consumed. As noted in the 1924 report of the Portland Cement Association, cement is

> a bulky commodity [so] it is not profitable to ship it far from the mill, and since the limestone and coal that are the chief materials used in its manufacture are widely distributed throughout the United States conditions have been favorable for the establishment of cement mills in nearly every state.
>
> Each cement mill thus enjoys a local monopoly, the radius of which is determined by its cost of production and by the proximity to other mills. The cement mills of the South, which possess an abundant supply of a limestone almost ideal for cement making, joined to a cheap supply of coal, can produce cheaper and send their product farther than the mills near New York City, which possess fewer advantages in respect to raw materials.[44]

Times have changed, obviously. While it is true that for a long time cement was produced by relatively small installations, in the last several decades cement giants have developed as small concerns were gobbled up by bigger ones trading in larger markets. McInnis Cement, even though it is the largest cement plant in Canada, is bucking the trend.

The cement my dad forbade us from playing with probably came from one of the several cement plants in Southern California. While a small natural-cement plant briefly operated southeast of San Diego in the 1880s,[45] serious production on the West Coast didn't begin there until 1898, when a plant was opened in Colton, about eighty kilometres east of Los Angeles.[46] Over the next ten years, plants were built in the San Francisco Bay area, near Portland, Oregon, and notably, in Concrete, Washington.

The summer after I visited Port Daniel, my family and I went through the last town situated on the Baker River, a stream rising in the high country of the North Cascades mountains in Washington State. Limestone was abundant, as were deposits of clay that contained the silica and alumina compounds that work so well when fired to make cement. A history of the town says that cement-making began shortly before the turn of the twentieth century, but grew exponentially after World War I. The home-grown product was used to build the Lower Baker Dam in the 1920s, which for a time was the highest hydroelectric dam in the world at eighty-three metres. While it seems almost miniscule today, when there are a dozen hydro dams towering three times as high, for decades it was an impressive advertisement for the quality of the cement created in the town and shipped all over the Pacific Northwest.[47] One of its biggest jobs was the building of Grand Coulee Dam in eastern Washington, about which we'll learn much more in the chapter on water.

Some limestone deposits seem almost inexhaustible. One of the oldest limestone quarries in North America began production near Quebec City in 1651. Records show that by 1670 quantities of stone were being quarried there for building houses in the capital of Nouvelle France as well as for making lime mortar.[48] Since then, it has been in operation almost continuously and now supplies limestone to Ciment Québec, one of McInnis Cement's competitors, for making Portland cement.

But the raw materials of cement and concrete are not, in fact, infinite. Despite what initially seemed limitless deposits, the cement plant at Concrete shut down in the 1970s because easily exploited rock had been exhausted. In Southern California, the California Portland Cement Company plant at Colton lasted until 2013, by which time Mount Slover, the source of its limestone, had been blasted to hill status.[49] About 100 million tons of rocks were removed, reducing the 275-metre-high promontory to ninety-one metres. At least a dozen other cement plants in the United States and Canada have closed since the start of the twenty-first century. Reasons given usually include exhaustion of easily quarried limestone, inefficiency of the plant itself (particularly when faced with stricter environ-

Figure 2.4: Limestone is crushed in preparation for making cement at the McInnis plant. Photo: Mary Soderstrom.

mental controls), and general economic woes following the 2008 Great Recession.

Certainly, due to changes in the industry the price of cement has dropped substantially since the beginning of the "Cement Age" in 1916, a barrel of cement cost US$1.56, according to U.S. government economic statistics. That would work out to about $34.35 in 2016 U.S. dollars, but in 2016 the actual price per barrel was about US$18.94.[50]

Making money in the cement and concrete industries depends on many things, among them the demand for the product and how cost efficient the production process is. McInnis and its corporate and public investors are banking on demand for cement being strong, and that its new kiln will produce cement at a lower cost than its competitors. Indeed, McInnis bills itself as the most efficient, least polluting cement plant in North America, ready to take up the slack in the supply of cement anywhere in the eastern half of the United States or Canada. But note that, while it's big for a North American plant—the biggest in Canada with its production of 2.2 million tons of cement a year in one production line—there exist bigger cement plants in several countries. Two in South Korea produce four million and ten million tons a year respectively, even though the country has little limestone and must bring in the rock from as far away as the United Arab Emirates.[51] McInnis Cement is also miniscule compared to the biggest cement companies, which have their home bases in China and Europe, but more about that later.

At Port Daniel the limestone to be used is blasted from the hill behind the plant. When I was there, the limestone face was very close to the plant, but as production proceeds, the hillside will retreat backward as the rock is removed. How long this supply will last is not clear: preliminary studies suggest that there's enough good quality limestone in the immediate surroundings of the plant for half a century of production, and that somewhat less perfect limestone on the other side of the 132 highway (the main thoroughfare on the Gaspé Peninsula) could last another fifty years.

Is that enough to make the c$1.45-billion investment—most of it public money—in the cement plant worthwhile? Back in the mid-1990s, the idea of a cement plant half the size was first floated, but the entrepreneur wasn't able to get financing. Then, in 2011, that firm and the construction permits it had acquired were purchased by Beaudier, a concern controlled by the influential family behind Bombardier Inc., the company that began by making snowmobiles and now makes airplanes and equipment for public transport. Beaudier revived the idea and in 2014 persuaded the Quebec government to put up c$250 million in the form of a guaranteed loan. The Quebec pension fund, the Caisse de dépôt et de placement, chipped in c$100 million, while Beaudier and other private investors put up c$750 million. Concerns about how polluting a new plant would be were raised, which may be why McInnis today stresses that the plant was designed to comply with the National Emission Standards for Hazardous Air Pollutants (NESHAP) that went into effect in the United States in 2015, even though technically it doesn't have to. In the end, however, it wasn't the "green" aspect of the project that presented the biggest stumbling block but the enormous cost overruns once construction got started. By 2016, construction costs had ballooned by c$445 million. That's when the Caisse forced out the sitting CEO, put in another c$250 million, brought in another c$125 million in private financing, and became the major shareholder in the project.[52]

McInnis officials insist that the prospects for the new cement plant are good. Cement will be shipped to Canadian markets, but a large part of production will be picked up by the U.S., where cement consumption has been greater than its production for more than a decade: in 2018 the difference was about ten million tons. At least a dozen inefficient, more polluting U.S. plants closed following 2008, and these are unlikely to ever reopen because upgrading these plants to meet NESHAP standards would make them uneconomic. Barring a major rise of the Canadian dollar in relation to the U.S. dollar, and supposing the Trump administration doesn't slap special tariffs on the cement industry, McInnis should do very well, I was assured.

Yet, even when I was there, an invisible cloud of doubt hung over the spanking new facilities.

THE BEAUTY OF THE VARIOUS KINDS OF CRAP

Back when I started my trip along this concrete road, I sat down with a friend of a friend, an engineer who had worked on some big projects over the years. One of his first jobs was putting in a new railroad line in northern Canada in the late 1960s, a vast project through country filled with bogs and lakes. Getting the grading right was essential, and required moving a lot of earth from one place to another. "That's when I discovered I like dirt work," he said. He was ready to go on, to tell me about his other projects, but I stopped him there.

"Dirt work?" I said. "That sounds like games for big boys!"

He grinned, because I'd discovered part of the appeal of this kind of work.

He's not alone. I got a hint of how many others there are the winter after my visit to Port Daniel, when I attended the World of Concrete trade show in Las Vegas. It was overwhelming—58,000 registered participants and more than 1,500 exhibitors from all over the world. At any moment, hundreds if not thousands of men were gazing raptly at and talking excitedly about the latest in earthmovers and mud-makers. Portable, self-mounting concrete batch plants for use in making concrete on remote locations were bigger than a small apartment building. They were circled by admirers, some of whom were ready to put out really, truly big bucks to acquire one. Trucks with tires as tall as my two-storey house and designed to haul limestone to batch plants over rough terrain had their fans too. Outside, pump cranes too big for the exhibition halls stood high above the gawking crowds. It was a triumphant scene for lovers of "mud" and "dirt."

Furthermore, as I wandered around and sat in on seminars, it became abundantly clear that modern concrete benefits from many other things that some folks consider only fit for landfill: several kinds of industrial waste are proving to produce cheaper, and in some cases more environmentally friendly, concrete than the standard Portland stuff. In fact, as conference speaker Rick Scezcy told about 300 construction types at one of the sessions I attended, "If you got a pile of crap, someone is going to say, let's try it in concrete and see what happens." He went on to list the waste products that are frequently used today: flue or fly ash from the coal-fired electric power plants, ground-up blast furnace slag from iron smelting, and silica fume derived from the silicon production process.

One he didn't mention was calcined rice husks, an additive that so far hasn't been used much in North America and Europe, but which is on the radar in countries where rice is the major diet staple.[53] It's the silica in the rice that does the trick, which is perhaps why the Chinese used sticky rice as an additive in ancient times. Calcined rice husk can vary greatly in quality, however, so experiments are being undertaken in places as widely different as Greece and Brazil to standardize the material.

Slag from steel mills—as opposed to slag from iron foundries—hasn't been used for concrete, and millions of tons encumber the landscape near steel mills. That may change in the near future as a technology is developed that uses the by-product to make a concrete-like mix that doesn't cure in the usual way hydraulic cement does. Instead of taking place in the open air, the hydration is accomplished in a chamber filled with CO_2, where the concrete sops up the greenhouse gas. The result is actually "carbon negative," says chief partner Chris Stern of Carbicrete, a Canadian start-up that has high hopes for revolutionizing the world one precast concrete element at a time.[54]

SAND

Volcanic sand was the key to making Roman concrete, as we've seen, and sand is still important for modern concrete, but in quite different ways. Sand, gravel, and other aggregate give bulk to the mix of cement and water that will become today's "Rock of Ages." Without them, cement is merely a hard substance that will crack easily once it's set, of little use for building roads or structures.

You might think that finding sand would not be a problem today, unlike in Roman days when volcanic sand was so important. After all, there's sand everywhere along the sea coasts, and in my childhood memories, I seemed continually to have shoes filled with it. That's because the house on which my dad spent so much effort do-it-your-selfing was not more than a few kilometres from the beach, definitely a walkable distance. The long, lazy summer I was fifteen (the last one before I started getting summer jobs), my friends and I would meet shortly after lunch on the street that ran directly toward the ocean, and then saunter beachward. At low tide the beach was wide, and we could walk on the packed sand for what I remember as long distances. The summer fog would be burning off by the time we arrived, and the upper reaches of the beach would bristle with surfboards. By the middle of the afternoon, the sun glared off the water, and there'd be sand inside our bathing suits, in our hair, in the snacks that we'd brought with us. Sand, sand, sand: when I got home my mother would make me wash off outside with the hose as she fought her losing battle against it being tracked into the house.

The beach itself had been there for probably several thousand years.[55] It lay south of the flood plain of the San Diego River, which in ordinary Southern California years did not carry much water. But the beaches existed because of the sand and silt washed down from the hills that form the interior of Southern California. In the past, the river had gone on rampages, shifting course several times in the nineteenth century, once even turning abruptly south and cutting off Point Loma, the peninsula which shelters San Diego Bay, instead of running to its north. This is the kind of behaviour that in other rivers has prompted massive efforts to tame the water's course: both the Colorado River in the American Southwest and China's Yellow River are cases in point, as we'll see in chapter 5, "Water." The action taken in San Diego was not dramatic, but long before I was a teenager, the river had been subdued by the El Capitan Dam, which impounded run-off and fed water though a concrete pipeline to the city's water system. The river's course had also been stabilized: it now ran through concrete channels, and concrete breakwaters kept it from changing course and affecting the increasingly important military and shipping harbour to the south. The re-engineering of the landscape included two large jetties made of concrete rip rap that protected the entrance to the northern bay, which had been transformed into a yacht harbour and recreational park.

I did not know it then, but I'd encounter the effects of concrete on rivers, wetlands, and beaches again and again as I explored the world of this extraordinary material. Nor did I know that the first dam on the river, built in about 1803 by Franciscan fathers who had started a mission in San Diego, was made with stone and hydraulic cement produced in kilns on the mission land. It was a technique introduced from Spain, descriptions now say; certainly it would seem to be linked to the tradition of handing down Roman ideas over the ages that we've already mentioned.[56]

This taming of the landscape had consequences that weren't at first apparent. Among them was the fact that the breakwaters at the entrance to Mission Bay, as the newly controlled waterscape was called, interfered with the prevailing currents along the coast. Before the jetties, the currents swept sand southward, toward Ocean Beach, keeping it wide and inviting. But by the time we were hanging out at the beach, the sand was not being replenished naturally. The new configuration also affected the mouth of the channel, which began to silt up. Dredging was necessary, and I remember big concrete pipes that discharged the dredged-up sand south of the jetty.

As I think back now, I note that my dad never used beach sand for his projects. I don't imagine there were restrictions on doing that: we collected driftwood and shells after storms, both of which have long since become more or less forbidden. More likely he thought that beach sand, unless carefully washed, wouldn't make good concrete, either because he'd read something somewhere, or because he'd learned the hard way.

In fact, if it's cleaned properly, beach sand is a lot better than desert sand for making concrete because it is spikier than the wind-polished grains of the desert. The microscopic lattice that grows as cement sets adheres better to the rough grains, making stronger concrete. But ordinary dirt must be purged from beach sand before it's mixed with cement and any other aggregate. The Romans knew this, and studies of ancient sites have shown that, as time went on, they grew more and more skilled at cleaning sand.[57] If the sand does contain salt, the concrete made with it can deteriorate rapidly, particularly if it is reinforced with steel. In fact, thirty-one construction firms in China's Shenzhen's province had their licenses suspended in 2013 because they were using sea sand to make concrete for building high-rise housing and it was feared that the salt-steel-cement combination meant grave structural problems ahead.[58]

Dredging up sand to augment the beaches of my childhood was one thing; importing much larger quantities of sand has now become essential for maintaining beachfronts along the sea coasts of the world. Rising sea levels—caused by climate change, remember—only make more intense the normal ebb and flow of ocean currents carrying sand. Shorelines scoured by increasingly violent storms require imports of sand to stay attractive. In addition, wealthy but geographically small places are making new land by adding sand to their shores. The United Arab Emirates imported hundreds of millions of dollars' worth of sand from as far away as Australia to build new developments literally from nothing,[59] while the city-state of Singapore is now 22 percent bigger in land size than it was in the 1950s because of imported sand dumped into offshore waters.[60]

The collateral damage from these grandiose developments by oil-rich countries and tiny but powerful states includes a growing shortage of sand that is made more acute by the rapid economic development of what used to be called the Third World. Sand may seem abundant, but in fact it is getting harder to come by. What some call the Sand Mafia has been scouring beaches in North Africa, the Maldives' islands, India, Cambodia, and Thailand to provide sand for the hungry construction market. Environmental damage to coastlines and riverbeds has prompted some countries to ban or strictly regulate it. In 2012, the Chinese government set out seemingly strict guidelines for sand mining in that vast country, but five years later a group of Chinese scientists appealed in *Nature* for a stop to marauding sand-mining ships scooping up the Yangtze river bottom.[61] In India, a plan for sustainable sand mining was proposed in 2016 but seems to be unenforced.[62] After Sri Lanka's long civil war ended in 2009, reconstruction projects began scooping up sand from rivers, imperilling water supplies. By 2017, court cases instituted by the government stopped the process in several rivers and along the coastlines.[63]

Some forty-six billion tons of sand are consumed every year in construction, and frequently these days, it and its cousin gravel can no longer be found near construction sites. This means that concrete makers must often look farther afield to find supplies within distances that make sense economically. There's semi-serious talk about shipping sand left by melting glaciers in Greenland to construction sites around the world.[64] Already, builders in Southern California have begun barging sand about 1,600 kilometres from quarries on Vancouver Island in British Columbia to build apartments. Yes, there was sand available much nearer geographically, but it turned out to be cheaper to use the imported stuff. Transportation costs were less for sea travel than for moving sand by truck from sand mines in the hinterland using Southern California's clogged freeway system.[65] Environmental concerns have also affected sand's price and availability in North America and Europe as regulations about where and when sand can be taken have been put in place and, what's more, enforced.

So, perhaps not surprisingly, the construction industry has begun talking about "manufactured sand," or m-sand. M-sand is basically rock or rubble that is crushed in huge machines, graded according to size, and then used the way naturally occurring sand and gravel are. At the moment its cost is frequently more than river or sea sand, in part because so much ordinary sand is taken with little or no payment from coastlines of countries where there is little or no regulation.

One Sri Lankan company, however, boasts about getting a certificate of excellence for the way it produces sand by crushing rock, producing a kind of sand that is perfect for building, and of course more environmentally friendly.[66] How valid its claims are is hard to judge from this distance, but what is true is that concrete makers all over the world are contemplating using more and more "manufactured" sand and other aggregate to make their concrete.

OTHER THINGS FROM THE EARTH

We've already mentioned slag from steel and iron smelting as other kinds of "dirt" that have been successfully tried as additions to concrete. But there's another sort of material that also has its origins in the earth and is perhaps even more important than the others because of how it literally reinforces concrete: the metal products made from iron ore.

Smelting goes back a long way in human history, to well before the Romans. They—and the Greeks before them—recognized the importance of metalworking and assigned a god to look over the hot, difficult skill. Perhaps not surprisingly, Vulcan—called Hephaestus by the Greeks—was also associated with the hottest, most powerful, and seemingly impetuous phenomenon the ancients knew: volcanoes. But while the Romans perfected the use of volcanic sand in their concrete, they did not systematically use iron with concrete. A few hooks and braces show up in their buildings, but it wasn't until the middle of the nineteenth century that what seems today to be a natural association of iron, steel, and concrete emerged.

Understanding why the partnership works so well requires a little thought about the particular strength of concrete.

Concrete can bear a lot of weight. Today, its strength is measured in either pound-force per square inch (PSI) or megapascals (mPa), that is, how much weight a given amount of concrete can hold without being crushed. The standard for road construction, for example, is frequently 4,000 PSI or nearly 30 mPa. Picture an SUV somehow balanced on a square inch of concrete paving without it collapsing and you have an idea of how strong concrete must be to meet the standard.

Figure 2.5: Concrete is pumped into place on a street project in Montreal. Photo: Mary Soderstrom.

But while concrete has great *compressive* strength, it doesn't have as great *tensile* strength. That's because, while the lattice of cement that glues the aggregate in the concrete together doesn't compress easily, it breaks when the latticework is bent up or down or laterally. Consequently, concrete is not as good at bearing loads that stretch it. It's a bit like your body: if you carry a load on your head (and in much of the world that's still a widespread method), you can carry kilos and kilos without being squashed. But if you take that same load, divide it in half, and try to carry the two halves in baskets at the end of your extended arms, you'll find that it's much harder to do, and you may not be able to do it for very long.

But what if you were to get some sort of support for your arms? They won't get tired so quickly; in fact, many societies discovered that you could carry much more if the buckets or whatever were suspended from a yoke or a pole that goes over your shoulders. In a way, reinforced concrete works in a similar fashion, with steel reinforcement acting the way extra support for your arms does.

Once modern concrete was invented and its possibilities began to be seen, you'd think the search would be on to find ways to increase the material's tensile strength. Steel or iron would seem to be obvious things to try. After all, metal frameworks had been used from the end of the eighteenth century to reinforce stone and masonry buildings: the Panthéon in Paris, built with stone and whose design was inspired by the Roman Pantheon, used just such techniques.[67]

Two considerations stood in the way, however. The first involved questioning by engineers and architects regarding whether the concrete would adhere to the steel reinforcement. Indeed, it took a bit of experimenting before the right configuration of reinforcing rods and concrete was found. The second was the relative scarcity of steel: it wasn't until thirty years after the first modern concrete constructions that the Bessemer furnace was invented, dramatically increasing the quality and the availability of the metal.[68] After that, the possibilities of steel construction were explored relatively quickly: the Eiffel Tower, completed in 1889, was considered the acme of metalwork. At 300 metres, it was higher than any church steeple in the world and dwarfed the buildings that surrounded it. (For those who lament current architecture, it's good to remember that when it was constructed, contemporaries called it ugly; obviously, tastes change.)

But it took decades more before reinforced concrete became commonly used for housing and monuments. One reason for this was a sort of snobbism. Concrete, particularly the reinforced kind, was not considered "noble" enough by many architects. Often it was used to build utilitarian things like bridges and roads. Worse in the eyes of some, one of its early promoters, Jacques Monier, was a horticulturalist and reinforced the big flower pots he made with wire armatures. The reaction appears to have been along the lines of: "*Vous n'êtes pas serieux!*" "You can't be serious!" While the use fits very well with the

Figure 2.6: Do-it-yourself building with concrete and rebar in the Amazon basin of Peru. Photo: Mary Soderstrom.

idea that concrete is closely tied to the ancients' element of earth, Monier's double career just made the case stronger among some that concrete was not fitting for buildings of consequence . . . or even for housing ordinary folk.

Nevertheless, the advent of steel-framed structures and reinforced concrete opened the door for construction of tall buildings, with all that implied. Before then, six storeys was about the highest that any building for housing or commerce could be built. Lovers of musical theatre will remember the rube in *Oklahoma* who sings his astonishment at seeing a building in Kansas City that's seven storeys high. "What's next?" his friends ask wonderingly. The answer, in the form of skyscrapers, lay not far in the future.

The reason for a height limit for masonry, stone, and wood structures is that, without special support, they reach the limits of stability at six storeys.[69] That's a major reason why Baron Georges-Étienne

Haussmann, the architect of modern Paris, decreed that the city's streets would be lined with buildings that height, and why, even back in Roman times, apartment blocks were no higher. Furthermore, before the invention of the elevator, it was hard work to get loads any higher. Anyone who's lugged a sack of provisions to an Airbnb rental on the top floor of a Parisian apartment house would concur.

Because of the Haussmannian height limitations, when the first reinforced concrete building was constructed in Paris, it was no higher than that, but when it was built in 1893, it proved to doubters that the material could lend itself to great beauty. That was exactly what architect François Hennebique had in mind when he built it to house his offices. The building, still standing at 1 rue Danton in Paris, advertised how well his patented methods for building with reinforced concrete worked. They also demonstrated how they could create the latest architectural style, including the curvy Art Nouveau bas reliefs and mouldings that were "the thing" then.[70]

The first reinforced concrete skyscraper was the sixteen-storey Ingalls building in Cincinnati, Ohio, built in 1903 and considered an engineering feat at the time. It used twisted steel bars inside concrete slabs to improve the adhesion of concrete to the steel—the ancestor of today's spiral rebar whose spikes you can see rising from many modern construction sites. Slabs, beams, and joists were cast as units in order to provide a more rigid and stronger structure, as well as to cut costs.[71] Not only was the resulting building cheaper to construct, but also it was more fireproof than steel-frame structures, which could melt in intense conflagrations.

The two technologies—steel framing and reinforced concrete—opened up new possibilities in construction but also introduced complications, since the new materials required careful calculations. It might be that pouring concrete required less training than doing stonework or laying brick, but new competencies were needed. Building became a cooperative venture, where civil engineers were extremely important, transforming the dreams of architects into sometimes breathtaking constructions. In Europe, the business model was one of licensing patents and hiring out engineering expertise. In North America, the consulting engineer and architectural firm became the norm.

In the middle of the twentieth century two other ways to use steel in concrete became common, opening the door for even more extravagant projects.[72] Instead of steel reinforcement passively strengthening the building simply by being part of the structure, the steel is stretched through a process that ratchets up the tensile strength of the reinforced concrete. This can be done by prestressing. In this case, high strength steel tendons are strung between two abutments and stretched to 70 to 80 percent of their ultimate strength. Then the concrete is poured around them and allowed to cure. When it has hardened, the stretching is released and the steel tries to regain its original shape, squeezing the concrete and making it stronger because it is being compressed. Poles, bridge girders, wall panels, and sometimes roof slabs are typically made this way.

Or the concrete can be poststressed. An example is the way the modules of Habitat 67, Moshe Safdie's groundbreaking construction for Montreal's Man and His World Exposition, are attached to each other.[73] In this method, concrete is cast around, but not in contact with, unstretched steel cables, which sometimes will be passed through ducts after the concrete has hardened. Then steel tendons are inserted and stretched against the ends of the unit and anchored externally, placing the concrete into compression. This method is often used for bridges, large girders, floor slabs, and shells.[74] At Habitat 67, the attachment points for the steel tendons can still be seen, but more about Safdie and his ideas later.

The end result of both these techniques is to allow architects and engineers to build lighter and thinner concrete structures that are still very strong, creating the skylines everywhere that are punctuated by higher and higher buildings. It was an example of a material born of earth entering the realm of the other elements.

CHAPTER 3
FIRE

CONCRETE IS BORN IN FIRE

As I said earlier, down at the dock the day I was in Port Daniel, an ocean-going vessel was offloading petroleum coke, the fuel for transforming limestone into cement. Pet coke is the residue from petroleum distillation. It is a material that for a long time was considered one of the side products of the fossil fuel industry that were more nuisance than anything else. It's tempting to say that this is another form of "dirt" that goes into making cement. Its utility for firing the huge kilns needed to calcine limestone was not at first recognized. The fact that a positive use has been found for it leads some apologists for the cement and concrete industries to say, "Hey! Look how we're recycling a substance that would otherwise be used for nothing but landfill."

Volcanic sand, as we've seen, was the key discovery that allowed the Romans to make such excellent concrete. The trass, which northern Europeans found was a good substitute, was also forged in the fires of the earth. And once curious engineers began playing around with ways to make modern concrete, their essential breakthrough came when they replaced the immense heat of volcanoes with fire harnessed in kilns, where the basic chemistry of the materials was transformed in the great heat. Therein lies both the magic of modern concrete and the menace it wields, which threatens the health of the planet itself.

Before there was cement there was lime, and making it had a large impact on the forests of any region where people had figured out how to make the substance. Making Roman concrete was even more energy devouring. Robert Courland, in his *Concrete Planet*, imagines a discussion between Agrippa, Herod the Great, and his chief engineers about building those harbour facilities at Caesarea Maritima on the coast of what's now Israel, mentioned earlier. In addition to all the pozzolans required for the hydraulic cement, a whole lot of wood was going to be needed for firing the lime kilns. One oak tree is needed to fuel a kiln to produce 190 kilograms of lime, so between 100,000 and 200,000 trees would have to be sacrificed for the breakwater and port. The engineer suggests that the decimated forests of the Middle East will be insufficient for the task and so the making of the lime should be outsourced to the banks of the Danube. There the finished lime could be packed into wooden barrels and then shipped down to the Black Sea and to the Mediterranean through the Bosporus strait.[1]

Of course, we'll never know if that kind of discussion ever took place, or even exactly where the lime used to make Roman concrete for the project came from, but the long shadow of deforestation—which of course ironically involves eliminating those shady green forests—falls darkly over the Middle East from the time that people started living in cities.

Certainly the forests nearest the limestone quarries went first. More recent examples abound. On the island of Montreal, for example, many of the forests fell long before the *colons* were ready to farm the land on which they grew, because the wood was needed for cooking, heating, and making lime to cement together the building stone used to make houses in this cold climate.[2]

Wood, however, has been used in the building of housing for thousands of years. Even in regions where trees appear never to have grown in abundance, like the great area between the Tigris and Euphrates rivers, timbers were used as doorjambs and in stairs or ladders in dwellings built with sun-dried brick.[3] More elaborate buildings required more wood: the forests where the massive cedars of Lebanon grew stretched "10,000 leagues in every direction" according to the oldest text extant, the *Epic of Gilgamesh*, dating from 2600 BCE. They supplied builders of the eastern Mediterranean, Mesopotamia, and even Egypt with lumber for hundreds, even thousands of years.[4] But by Roman times their scarcity inspired the Emperor Hadrian to place what was left under his personal care in 130 CE.

Even before then, though, the Romans had begun to look to the forests of the north, which seemed as endless as those of the cedars of Lebanon once appeared. Pliny the Elder, writing in the first century BCE, noted that the wood required for the business of living was being brought from farther and farther from Rome. Wood was fuel for baking bread, for heating water for the Roman baths, and for warming lodgings; texts exist exhorting landowners to plant trees and set aside woodlots for these purposes. How effective this was is unclear, even though in places blessed with enough rainfall, wood for many of these activities could be supplied locally if the forests were skilfully managed: the careful harvesting of shoots from coppiced trees in the British Isles is one example.[5]

The fact that trees will regrow is important today too. It's the basis of the assertion that using wood waste and other kinds of organic material—also called biomass—as fuel to fire cement factories could make them "carbon neutral." The reasoning is that because trees and other plants take up CO_2 as they grow and release roughly the same amount when they're burned, the CO_2 equation is more or less equal. McInnis Cement all along has proudly said that its facilities are designed to burn both wood from Gaspé forests as well as pet coke. Initial plans were for wood and other biomass to be used after eighteen or twenty-four months of operation, ramping up to about 30 percent of the fuel when the roll-out was completed. By late 2019, however, a feasibility study was still underway, but McInnis spokesman Maryse Tremblay repeated that the company still has that as an objective.[6] The baseline for measuring carbon emissions—which McInnis Cement will have to pay for in carbon credits if it doesn't reduce greenhouse gas emissions, under Quebec's cap and trade system—was 2017, the first year of production, when pet coke alone was used. So starting off with a fuel that produces a lot of CO_2 and then switching to a fuel that is supposedly carbon neutral should be good for the bottom line. (It should be noted, though, that CO_2 from the fires that cook limestone contribute about only 40 percent of the CO_2 produced in making cement: the rest comes from the chemical process that turns limestone into cement, and as such can't really be reduced.)[7]

But is wood as fuel really carbon neutral? The question is a big one and has to be considered when thinking about concrete's effect on the world. Engineers at McInnis told me that it's usual practice to assume that over twenty years new, replacement timber growth will take up the amount of CO_2 equal to that freed when a tree is burned. But the assumption is based on quick-growing, managed softwood forests, so it might not be applicable in Gaspé, where it takes longer for trees to grow and reforestation is not guaranteed. Other scientists suggest that the assumption doesn't work even in places like the southeast United States. That is because there can be differences in the carbon uptake among various tree species, as well as damage done to the forest soil during the time the land is denuded of trees, leading to even more CO_2 release.[8]

Pet coke is supposed to be a cleaner fuel than coal, which had been the go-to fuel when the modern cement industry began. It's also increasingly used to generate electricity in countries like India, largely because it's cheaper at the moment than coal.[9] Its use in cement plants there is not viewed as being as polluting as coal. In setting up new pollution standards to fight the growing pollution problems of cities like Delhi, cement plants got a pass because, it is said, the particulate matter is contained within the cement plants, unlike the case of electricity-generating plants. But more about that in chapter 5, "Air."

A small amount of energy used in cement plants comes from alternate fuels, and at least some governments are pushing for recycled waste to become major sources. These include used tires, rags, and carpets, even discarded coffee pods from espresso machines.[10] But the ecological arguments in favour of recycling some forms of waste haven't convinced everyone of their appropriateness. For example, a plan to burn tires at a Brookfield, Nova Scotia, cement plant owned by cement giant LafargeHolcim has been the object of much local protest, including

Figure 3.1: William E. Ward's reinforced concrete house, built between 1873 and 1876, is the oldest extant structure of its kind in the United States. Photo: Daniel Case, Creative Commons Attribution 2.0 Generic licence.

legal challenges. Nevertheless, the way was opened in March 2018 for a year-long trial of burning tires when a Nova Scotia Supreme Court judge rejected a bid to block it. The trial will run into 2020.[11]

But nowhere in the list is there a truly renewable energy source to fire the great cement kilns. That is because, for good or ill, the end product of hydro, solar, nuclear, tidal, and wind power is electricity, and while electricity is necessary to power machinery, computers, and the like used in preparing the raw materials for the kilns and for administering the plant, so far it doesn't produce temperatures high enough to make cement from the basic rock.

Under ordinary conditions concrete doesn't burn, and in the nineteenth century this was one of the great arguments in favour of its use. Some of the first grand concrete constructions were showcases for concrete's apparent impermeability to fire. Examples are theatres and an imitation manor house in New York State, called locally Ward's Castle.[12] The latter is the oldest reinforced concrete building still standing in the United States. Built by a prominent industrialist, William E. Ward, between 1873 and 1876, the seventeen-room house was designed to show off the fine quality of concrete construction, but also to calm the fears of Ward's mother, who was morbidly afraid of fire.

This is not to say that concrete isn't transformed by fire. Those that followed the terrorist attacks on the World Trade Center on September 11, 2001, saw just how damaging a great conflagration can be to a steel-frame structure even when a large part of a building and its foundations are made of concrete. Inflammable contents burned and the steel was weakened from the heat of the fire: the result was the disastrous collapse of the buildings. Similarly, as we will see a little later on, the immense heat of nuclear reactors can turn concrete into a magma-like substance, melting it like volcanoes can melt the rock deep within the earth.

Nor should we forget that there's fire at the heart of the chemical reactions that make cement bind with aggregate. Engineers, architects, and builders must deal with the heat produced as concrete cures, particularly when the mass is large. As we'll see later, it's estimated that if concrete pours at major American dam sites had not been cooled with circulating water or dry ice, the concrete at the Grand Coulee or Hoover dams might still be curing, nearly eighty years later.

THE FIRE NEXT TIME: CONCRETE AND WAR

The prophecy about the end of the world being "the fire next time" is hidden in a quote from the New Testament, and it scared me to my toes the first time I heard it.[13] I must have been seven, big enough to know that grown-ups sometimes said interesting things when the kids had gone to bed, although I didn't realize just how scary those things might be.

That night my parents had friends over for dinner and I was supposed to be asleep, but I'd crept out of my bedroom to listen to them.[14] The Korean War (that one, so long ago, not the one that

North Korean leader Kim Jong-un has threatened more than once in the last few years) was just beginning, and the grownups were speculating about what might come next. All of them had vivid memories of the last war, but none of them were terribly concerned about the men present having to serve. They all were fathers, and fathers were exempt from the post-World War ii call-up and draft. Good, they all agreed. But then talk turned to the atomic bomb, and that was a different story.

They'd seen the photos of destruction at Nagasaki and Hiroshima; they knew the Soviets had developed their own bomb. The question that floated in the air that night was: would it be used this time around?

That was when one of the women—probably a little drunk, maybe a little hysterical at the idea of another war—began to talk about "the fire next time." Bombs. Cities on fire. Children dying. The end of the world . . .

I must have cried out, because the next thing I remember I was being bundled up and hurried back to bed. Don't worry, my mother said. We'll be all right.

She didn't say that concrete would protect us, I'm sure, but over the years I've come to realize that concrete was part of her assurance. Air raid shelters, strong buildings: we had them, we would build more, we'd be safe from the fires of war. But, I see now, that just as fire is an integral part of the making of concrete, it also is the chief element in war, the scourge from which we now suppose concrete will protect us.

In earlier days of the material, however, the major use of concrete in war wasn't in cities. Rather it was used to construct roads, canals, bridges, and other infrastructure that allowed the generals to conduct war. Its use in fortifications was also essential from the time that military technology produced high calibre cannons: big walls were needed to protect from big guns.

It's no accident that Louis Vicat, the Frenchman whose pioneering experiments were essential to the development of modern concrete, worked for the French government's Departement de ponts et chaussées, where bridges and roads were planned and built, and which originally was a war department. French and German engineers were first and foremost military engineers: in North America the idea of civil engineering came about as a way to differentiate the use of engineering for civilian projects from engineering for war.[15]

Masonry forts combined with earthworks had been standard for centuries, but by the time of the American Civil War (1861–1865), it became clear that they were usually no match for being pounded by the more powerful guns being developed. In 1868, the New York Board of the U.S. Army Corps of Engineers laid out specifications that required using concrete in building weapon storehouses. By the mid-1880s, the French were systematically testing concrete fortifications by blasting concrete ramparts of varying thickness with the highest calibre guns available. Adding steel reinforcement to concrete increased protection, as the Belgians discovered to their dismay at the beginning of World War I when unreinforced concrete forts were overrun relatively easily.[16]

That war to end all wars was concrete's baptism of fire, so to speak. Not only were forts and guns firmly anchored in concrete and protected by the substance, but as the war straggled onward into a war of attrition, concrete also played a big role in the trenches.

Take the report of an American Rhodes Scholar who somewhat bizarrely was invited to visit the German front lines in the spring of 1915 when he was on holiday from his studies at Oxford. (You'll remember that while the Great War began in the summer of 1914 for European powers, the United States did not enter it until 1917.) F.H. Gailor's account was published March 24, 1915, in *London Daily Mail*, but the trenches he visited had been built the fall before. They "seemed permanent enough for spending many Winters . . . . [Some] have two stories, and at the back of many of them are subterranean rest houses built of concrete and connected with the trenches by passages. The rooms are about seven feet high and ten feet square, and above the ground all evidence of the work is concealed by green boughs and shrubbery."

He continues: "One officer described the life as entirely normal; another said, in speaking of a Louis xv couch which had been borrowed from a nearby château and was the pride of a regiment, 'Oh! we are cave-dwellers, but we have some of the luxuries of at least the nineteenth century.'"[17]

That's a far cry from other accounts of the trench warfare, but it gives an idea of how concrete could change some of the worst aspects of war. The Germans ultimately surrendered, of course, in large part because they had overextended themselves and were under pressure from a severe food blockade by the British Royal Navy. The terms of the end of the war were greatly disadvantageous to the Germans, which many say laid the foundation for the German lust for rearmament and conquest twenty-five years later. Certainly, during the interwar period much effort was expended by all sides to prepare for another war, which was expected to be fought on the same principles and with the same techniques as the Great War.

Chief among the French defences was the Maginot line, more than 700 kilometres of fortifications, tunnels, and obstacles along France's borders with Belgium, Luxembourg, Germany, Switzerland, and Italy. The defence was constructed between 1928 and 1940, and required millions of tons of concrete. The forty-four grand artillery emplacements, the sixty-two infantry bases, and thousands of smaller blockhouses were conceived as a unified defence system. Connected by tunnels where electric-powered railcars ran and with room for thousands of soldiers, the system wasn't completed when World War II broke out in 1939. But even the completed portions offered little defence since the Germans were able to get around the line by going through Holland and Belgium. They marched toward Paris in 1940 without serious opposition from the troops holding the line. Indeed, when the French government surrendered in June 1940, the 25,000 men in the Maginot installations couldn't believe it and didn't initially obey orders to turn the defences over to the Germans.[18]

The French weren't the only ones to try to protect themselves behind walls of concrete. Czechs attempted the same thing, with just about the same degree of success. They built a 230-kilometre-long line of concrete blockhouses that would each shelter about thirty soldiers. The government capitulated, though, giving up the Sudetenland in September 1938, and setting the stage for full war a year later when Hitler's forces marched into the country.[19]

At home in Germany, the Wehrmacht forces travelled by rail and along a system of highways that had no equal anywhere in the world.

During the interwar period the Germans put much effort into building the system. Somewhat ironically, however, the highways of the Autobahn were intended for recreational driving when the program was begun in 1933. Art historian Adrian Forty says that the assumption was that freight would move by rail so that highway driving would be only cars filled with people "reconnecting" with the land. "Driving on an autobahn would not only allow the urban dweller to enjoy the German landscape, but more particularly the chance to experience it in a wholly novel manner, unfamiliar even to the railway traveller." He quotes one engineer: "We must build not the shortest, but rather the most noble connection between two points."[20] Routes were laid out that gave marvellous views—of the Alps for example—and beauty was designed into highway bridges and filling stations.

When the National Socialists came to power, the strategic benefits of the elegantly planned highway system were not lost on Hitler and his associates, and the highway system proved to be extremely useful for the Nazis' war effort. It is no accident that after the war the United States launched an ambitious program of interstate highway construction when former general Dwight Eisenhower was elected president. As a young soldier in 1919, he had been part of a military convoy that took sixty-two days to cross the United States. Twenty-five years later, in contrast, he and many of his staff saw firsthand the utility of a good highway system in both Italy and Germany.

Authorized in 1952, the U.S. Interstate system was designed to incorporate existing good roads across the country, upgrade others, and build new ones. None of these roads could have been built without concrete. They also paved the way, literally and figuratively, for the great expansion of cities in the post-war period. Along the way, the idea of using highways to link people to nature—like so much of Nazi philosophy—has proved to be a complete bust. But more about that further on when we talk about concrete and home fires.

When the tide began to turn against them, the Germans turned to concrete again. At the time of the Japanese attack on Pearl Harbor on December 7, 1941, the Germans controlled most of Europe and had set their sights on the Soviet Union, opening up an Eastern Front. But the entry of fresh, new forces from the United States led the Germans to fear an attack by the British and their allies. Consequently, they divided their resources and began building the Atlantikwall on the Western Front while continuing to battle in the east. The line of concrete bunkers, anti-tank ditches, mines, blockhouses, and obstacles like kilometres upon kilometres of barbed wire stretched along the Atlantic coast from the French-Spanish border northwards to Norway. The amounts of concrete used were stupendous and were supplied by companies like Lafarge, the ancestor for today's giant, which openly collaborated with the Germans.[21]

Today, you can tour the gun emplacements and bunkers, walk along paths between fortifications, and imagine what it must have been like more than seventy-five years ago when war was raging. War tourism is a big industry, and has been since the world began to return to normal in the 1950s. Visits to bunkers are even possible in North America, which was barely brushed by conflict in either world war. My husband and I visited one of them on a summer day in the 1970s. It was on Point Loma, which guards San Diego Bay and which had been the background for much of my growing up.

The point had been home to military installations for nearly 200 years by then. The Spanish and after them the Mexicans built lighthouses on it and manned a small garrison for the first part of the nineteenth century. After the Americans' war with Mexico ended in 1848 and California became a state in 1850,[22] the southern part of the point was assigned to the U.S. Army and named Fort Rosecrans. The heart of military operations was on the lower levels on the bay side, not the ocean.[23]

But the day of our exploration we looked out to sea from near the high point on the peninsula. The morning fog had burned off, leaving behind the smell of damp sagebrush and wild fennel. The sun was now high in the sky, off to our left, to the south, glinting off the ocean. Up ahead was the Cabrillo National Monument, an homage to Juan Rodriguez Cabrillo (or João Rodrigues Cabrilhão, his birth name in Portugal), who captained the first European ship to enter the harbour, back in 1542.

That was eight years after Jacques Cartier ventured into Port Daniel and just about the time that Vitruvius was being rediscovered by engineers and architects interested in building in the Roman fashion. The California site had changed much more profoundly than had the Gaspé one, but there were concrete points of connection nevertheless.

One of them was concrete, in fact. There is a lot on Point Loma. That was clear that morning as we stumbled upon what had been the ultimate in defensive structures during World War II.

We turned off the street that ran along the spine of the point, down a side road that led toward the ocean. Nobody was out, no vehicle was parked along it—if I hadn't known better I would have thought this part of the military installation had been completely decommissioned. So there seemed no harm in getting out of the car to look around.

The first sign of the weapons was on the ground in front of us: a series of steel tracks embedded in concrete that were used for moving the big guns that were positioned there during World War II. For some reason, though, what they connoted didn't sink in at first. Instead we turned toward the hillock to our left, ready to scramble up to get a better vantage point from which to view the ocean. There were a few scrub oak there which are native to the area, but given the way they were grouped together, perhaps they'd been planted to camouflage the gun emplacement. I wasn't thinking of that because out in front of us was the ocean. It was so clear we could see the Coronado Islands, thirty-two kilometres to the southwest off the coast of Mexico: the air was as sparkling as it would be the day I visited McInnis Cement.

But closer to the edge we saw below us the concrete structure that must have once housed the guns. The concrete walls were about a metre thick and capable of withstanding a pretty strong bomb blast. They weren't unique, just one of myriad installations built along this coast and others to provide protection in time of war. They were useless now, and they had been for nearly a generation, ever since the advent of nuclear weapons.

The atom bomb marked the end of World War II, and with it came major changes in warfare. The fear that my parents' friends expressed at the beginning of the Korean War—the fire next time—was directly linked to this major paradigm and weapons shift. Just as the high calibre exploding shell made mincemeat of nineteenth-century fortifications, so nuclear weapons swept away earlier ideas about how to shelter both civilian populations and military forces. What could have withstood a direct hit by a conventional bomb was now practically useless. The new questions were: How to shoot down a bomber carrying an A-bomb? How to protect civilian populations during a war that could be over in a couple of hours, but whose effect would render the landscape unlivable? How to intercept a long-range ballistic missile designed to carry nuclear warheads?

Jump forward a few years from that expedition into the past of warfare. It was summer again, with car windows open, hot sun beating down, grain fields stretching in all directions. Not much traffic on the road: Sunday afternoon in North Dakota. We were on another road trip, having stepped away from all responsibility for a couple of weeks. We were camping most nights too, and that pushed the world even further away. The Cold War was on, but I didn't want to think about politics, foreign or domestic.

But then the F-15s, two of them, swept across the sky. Their presence seemed an intrusion from another reality, one that was the opposite of the bucolic landscape we were driving through. Appearances can be deceiving, though. What we didn't know was that just a little to the east was Minot Air Force Base, home to the jet fighters and also to a Minutemen missile installation. Nuclear weapons were stored there deep underground in concrete silos, ready to soar to the other side of the world at the turn of a specially coded key. There were many facilities like this all across North America. On the other side of the Arctic Circle, in the Soviet Union, there were just as many, if not more, all with missiles pointed toward targets in our neck of the woods. The silos were designed to withstand direct attacks by the enemy's bombs or missiles. So were the handful of shelters intended to protect military commanders and civilian leaders. None of them would have been possible without high-density, reinforced concrete designed to withstand a megaton atomic blast.

A megaton: that's what the Minuteman missiles carried. When I was a pre-teen in San Diego, the newspaper had published maps showing what would happen if a bomb that size was exploded over the harbour: our hillside houses would be more or less vaporized.[24] As another of my parents' friends said at that time: "Take cover? Are you crazy? I'm going to grab my wife and a bottle of booze and go out in style."

Nevertheless, air raid sirens blew for a minute or so at noon on Mondays, and there were occasional in-school drills to rehearse the protective measures we were supposed to take in the event of attack. For a while it was: duck under our desks, hunch forward on our knees and link our hands behind our necks. Just how that was to keep us safe wasn't clear even to fifth graders: I remember our teacher shifting the subject when I asked. An earlier plan had everybody heading out of school to neighbouring houses when an alert came, she said: her message was that this made more sense.

Maybe. But it became clear by the early 1960s that safeguarding the civilian populations in a nuclear war would require a lot more than that. During the world wars large cities had designated bomb shelters, and I suspect that was the kind of thing my mother had been thinking of when she reassured me back at the beginning of the Korean War. The Moscow Underground and the London Tube—both largely concrete structures—are two examples. My brother-in-law's mother had stories of bundling him up when he was a newborn and taking refuge deep beneath London in the Underground during the Blitz.

But by 1961 *Life* magazine was running stories on how to build your own fallout shelter, and please note the difference in the language used. It had become clear that nothing short of massive concrete constructions could actually protect a population from a direct hit by a nuclear weapon. Building such a network would be enormously expensive, if it were possible at all. So, under President John F. Kennedy, the idea of civil defence shifted. People were urged to build their own backyard shelters to protect themselves from the fallout that would follow a nuclear detonation. Nothing was said about withstanding a direct hit, nor was there mention of the fact that the shelters would be of little use in the cities that would be among the first targeted by missile attacks. Much, much more than a dugout protected by ordinary concrete or—heaven forefend!—sand and gravel would be needed when a nuclear war began.[25]

Many governments around the world did indeed strive to build effective secret shelters in which a relative handful of people—their leaders—would take cover in case of nuclear attack. There they would be protected from radiation and, in some cases, even a direct hit by metres of specially formulated, very dense concrete and/or mountains of rock and earth. The rationale for doing this was that for a nation to continue after nuclear war began, key players had to survive in order to coordinate a counter-attack and then to plan recovery. There are shelters under the White House in Washington, DC, and one in Florida, not far from where John F. Kennedy and his family vacationed at Palm Beach. In the late 1950s the United States also built a shelter big enough to house every member of the U.S. Senate and House of Representatives underneath the toney Greenbriar Resort in the mountains of West Virginia.[26] Officially called a conference facility, there was no hint of the resort's secret mission. Medical associations and car companies frequently held conventions and meetings there for more than thirty years without knowing that behind three foot concrete walls there was space for all 535 Congressmen and Senators to sleep and deliberate. North of the border, Canadians built more than a dozen bunkers for government and military personnel between 1959 and 1962: while not a big nation, it would be under the intercontinental ballistic missile flyway, should the United States and the USSR ratchet up the Cold War to something hotter.

In Canada, these Diefenbunkers (named after the prime minister of the day, John Diefenbaker) were decommissioned after the collapse of the Soviet Union and the end of the Cold War in the early 1990s. One of them, designed to shelter Canadian government officials, was closed down in 1994 after thirty-two years of service. It has since been opened as a museum and gives a glimpse of what life would have been like, had the Big One fallen.

Located about forty kilometres to the west of Ottawa, the nation's capital, the four-storey underground installation looks innocuous from the country roads that wander nearby. Cows still graze on top of it, remnants of the herd that a farmer named Billy Montgomery worked as the Cold War got hotter. Construction was top secret on what was still officially Montgomery's land, but word inevitably leaked out as crews began by excavating a quarry-like hole and then started pouring the first of 32,000 cubic yards of concrete. The concrete was high density and was poured by hand because it was too heavy for ordinary equipment to handle. Some 5,000 tons of steel were used, including reinforcing bars four inches in diameter: ordinary rebar is only about a third as robust. The concrete itself was 1.2 metres thick in places and used a mix that was 50 percent stronger than usual commercial grades. The shelter was designed to withstand a force of 160 pounds per square inch: were a five megaton bomb dropped 1.77 kilometres away, the structure would only shift, undamaged, four centimetres along the gravel bed on which it was constructed.[27]

When a reporter from the *Toronto Telegram* got wind of the work during construction, he hired a plane from the nearby airport at Carp and took a series of aerial photos. They showed massive construction underway, much more than could be expected from the official story that perhaps there might be a radio communications facility going in.

"Then why are there seventy-eight cardboard boxes containing toilets?" the reporter asked. He never got a satisfactory answer.

Clearly, the installation's existence was not a secret, although what was going on there was. This Diefenbunker employed hundreds of military personnel, each working thirty-day shifts in rotation. The place could house up to 535 people, the number of government members the facility was designed to protect, plus staff. No one, not even the prime minister or the governor general, was supposed to be allowed to bring in family members; those who were tapped to be survivors were those who would be necessary to keep a country running at a minimal rate.[28]

Bunk rooms had eight beds that were supposed to be used by sixteen people, sleeping in twelve-hour rotations. Only the prime minister and the governor general had single rooms; everyone else shared, but the ambience was designed to be upbeat. Chairs and other furniture were bright colours, and one wall of a common room featured a scene of river and mountains from the Rockies in order to open up the place a little. And the food was good: in a video prepared for the museum, former soldiers, now past retirement age, fondly remember how at midnight they could call up and order a T-bone steak. Because the shelter was designed to house people for thirty days after an attack, enough fresh food for a month was constantly restocked, which meant that the menu was varied in order to make sure that nothing was past its due date.

Vestiges of shelters like these exist around the world. China's massive ones are now used by students and sojourning workers as more or less temporary housing. In Albania thousands and thousands of small shelters put up during the reign of communist dictator Enver Hoxha remain. More concrete was used to build them than was used in the Maginot Line, and observers argue that the campaign, which was designed to keep Albania from being overrun by either the Soviets or NATO forces, seriously impeded the progress of the country because scarce resources went into them, not development. Now some of the shelters have become squats for people without housing, while others, according a BBC report, are places where lovers go for trysts in a country where there still isn't much private space.[29]

It may seem crazy today to have invested so much in shelters because it's easy to forget how imminent war seemed during thirteen or fourteen years of tension, one of the former soldiers who had been posted to the Diefenbunker near Ottawa says in the video. Certainly it was with a great sigh of relief that most of the world greeted the end of the Cold War.

Recently, though, signs are appearing that suggest nuclear preparedness is making a comeback. In Canada two sites in unnamed places where high-level government officials could shelter in extreme emergencies got the okay in the fall 2017 as North Korea began testing long-range missiles. A guide at the Diefenbunker museum grins when you ask if the new shelters will be like this one: "Couldn't say if I knew: it's a secret." That's similar to what PBS reporters learned about the Greenbriar Resort shelter. It probably would still be secret, had it not been exposed in 1992 by *The Washington Post*. You can be sure there's another one even more secret now, however, that would allow the United States to soldier on in case of attack. After all, Russian president Vladimir Putin has been bragging that Russia's new nuclear weapons can't be intercepted and North Korea is still governed by a mercurial leader who has a nuclear arsenal, while the same could be said of the United States, whose president has mused about blowing up Iran.[30] As for private citizens, some of them with pots of money are buying up missile silos turned into luxurious shelters against disaster and nuclear war. One near Las Vegas sold for near US$18 million in 2019.[31]

On the other hand, concrete has also proved essential to the development of nuclear energy for peaceful purposes. The elaborate structures shielding radiotherapy installations where thousands and thousands of cancer patients are treated every year would be impossible without high-density concrete. The hospital where a while ago I spent five mornings a week for five weeks getting blasted for a very early stage of breast cancer went out of its way to make the waiting area cheerful despite the shielding, and despite being partly buried in the side of a hill. The colours were warm reds, yellows, and oranges set off by white walls. In the middle was a huge multicoloured jigsaw puzzle that people were encouraged to work on while they waited. At the far end, a long way from the treatment rooms, was a skylight that let in some daylight. Yet the low light of the radiotherapy space and the way the technicians scurried behind the thick walls while the machines hummed out their beam of radiation at my left breast made it clear just how serious this business was. For an installation like this, tons of specially formulated concrete were necessary.

Another development that followed from the atom bomb was the use of nuclear reactors to make electricity—and concrete is essential to that too. When exploration of the atom and its power began at the turn of the twentieth century, scientists followed paths that would lead to either weapons or peaceful uses. After World War II ended, attention was turned back toward studying how the atom's great energy could be harnessed to produce heat and electricity. In 1953, President Dwight Eisenhower set in motion the United States' Atoms for Peace initiative, which was designed to make nuclear power a main energy source for the world.[32] Canada's Candu process using unenriched uranium and heavy water was developed in the late 1950s. The USSR opted even more heavily for using the atom for power generation. Since the 1970s, developing countries such as China have invested in nuclear power too: some thirty nuclear power plants were constructed in China during the first decade of the twenty-first century.

Standard nuclear reactors of whatever type use concrete to isolate their guts where controlled chain nuclear reactions take place. In most cases, the enormous heat generated by the reaction is used to boil water, making steam that turns turbines, just as heat from coal or natural gas is used to turn turbines in a conventional power plant. The same properties of high-density concrete that control radiation in cancer treatment centres or shield civil defence shelters also capture the gamma rays and other radiation created during the chain reaction. When something goes wrong, concrete and reinforcing steel are what are supposed to avoid a dangerous situation from developing.

And sometimes things can go disastrously wrong.[33]

Take for example the tragedy that began on a balmy night in late April 1986. Technicians at the No. 4 reactor in the Chernobyl power plant were starting to run tests on what would happen in case of a power outage.[34] The installation was about 100 kilometres north of Kiev in the Ukrainian Soviet Socialist Republic, part of the USSR at that time. Workers lived nearby in Pripyat, a dormitory town on the river of the same name. One of nine "atom towns" constructed in the USSR beginning in the 1970s, Pripyat is a northern city—a bit farther north than Regina—where fine weather in springtime is like a blessing.[35] The Orthodox Easter had been the weekend before, and the May 1 national holiday lay just around the corner. It was a night to keep the windows open, to think about when to plant what in the community garden, to go fishing perhaps, as survivors remember it. There was nothing in the stars to suggest that disaster was coming. No, Pripyat was a wonderful place: "From the first day we came to Pripyat, I never wanted to leave. It was paradise. Everywhere there were roses and fruit trees, we could fish in the river and pick mushrooms in the forest. It seemed the place had been created especially for us," one woman told *The Guardian* newspaper decades later.[36]

But on that fateful night, errors in the testing procedure resulted in a massive accident in which an enormous explosion blew the 500-ton concrete roof off the reactor building. When one of the operators—a survivor interviewed for a television documentary in 2006—went to investigate what had caused the explosion, all he saw was stars above his head. Fifty tons of nuclear fuel flew upward into the atmosphere, an amount ten times more than released by the atom bomb dropped on Hiroshima. Radioactive waste was spewed over kilometres of countryside. Some 135,000 people were evacuated from the region, but as many or more were conscripted from all over the Soviet Union to work on the clean-up and to build what the Soviets called the containment object. Eventually winds blew the radioactive debris as far as Sweden, but the main damage was concentrated at the reactor site, in the town of Pripyat, and the surrounding peaceful fields and forests which now make up what's called the Exclusion Zone.

As noted before, steel inside reinforced concrete will melt if a fire gets hot enough. Concrete itself may also begin to fail at temperatures above 800 degrees Celsius. At Chernobyl it was hotter than even that, and the combination of radioactive fuel plus superheated concrete made a slurry that could have resulted in a critical mass explosion which would have caused even more damage. But what the Soviet Union called heroes—and indeed many of the men were indeed heroic since they knew full well how dangerous the work was, and how important it would be in avoiding even worse disaster—successfully put out the fires and laid liquid nitrogen gas-bearing pipes under the reactor to cool down the great internal heat.

There followed an intense period of several months when more than 100,000 men worked to remove radioactive soil and other debris, and to build a concrete sarcophagus to encase the reactor. It was completed as winter set in by people—soldiers and conscripts from as far away as Kazakhstan, Estonia, and East Germany—who worked three shifts a day, but only for five to seven minutes at a time because of the radiation levels. In all they used 400,000 cubic metres of concrete and 7,300 tons of metal framework.[37]

Nobel laureate Svetlana Alexievich recorded the remembrances of some of them in her stunning book *Voices from Chernobyl: The Oral History of a Nuclear Disaster*. It gives a human face to the courage and the despair of the people who worked, lived, and loved around the power plant. It also gives a glimpse of how concrete was intimately involved in the clean-up. Ivan Aleksandrovich Lukashuk, a private, told her, "You have to serve the motherland! Serving—that's a big deal. I received: underwear, boots, cap, pants, belt, clothing sack. And off you go! They gave a dump truck. I moved concrete." He added, "There it was—and there it wasn't. We were young, unmarried. We didn't take any gas masks. There was one guy—he was older. He always wore his mask. But we didn't. The traffic guys didn't wear theirs. We were in the driver's cabin, but they were out in the radioactive dust eight hours a day."[38]

Viktor Latun remembered, "We built support structures: laundries, warehouses, tents. I was assigned to unload cement. What sort of cement and where from—no one checked that. They loaded it, we unloaded it. You spend a day shovelling that stuff and by the end only your teeth are showing. You're made of cement, of gray cement, and your special protective gear is too. You shake it off in the evening, and the next day you put it on again."[39]

Then there was the case of reactor operator Leonid Totunov, "who was the one on duty that night and he pressed the red accident button a few minute before the explosion," Captain Vladimir Petrovich Shved recalled. "They buried him at the Mytinskaya Cemetery, like they did the others. They insulated the coffin with foil. And then they poured half a meter of concrete on it, with a lead cover."[40]

But that was far from the end of the story. Ten years later, after the Soviet Union had fragmented into autonomous countries, it became clear that radiation was increasingly escaping from the shielding as the original concrete container fissured due to weathering. In the best-case scenario, the men who had built the first sarcophagus had estimated that it would last no more than thirty years.[41] Furthermore, inside steel support beams which had been dropped in place from helicopters were not attached securely to each other and had begun to shift. Somehow or other, a more secure structure was going to have to be constructed, one that would last for a minimum of 100 years and allow robots inside to dismantle the reactor itself.

By then the border between the Belarus and Ukraine autonomous republics ran just north of Chernobyl so that the greater part of the Exclusion Zone was in Belarus. The two countries, struggling to develop after the fall of the Soviet Union, were in no position to finance and build the kind of structure needed to shield the damaged reactor by themselves. So, through a rather remarkable international effort, the US$30 billion necessary was amassed from a mini-United Nations of Western countries and former Soviet bloc ones in order to construct what became the largest structure ever built and then moved. After nearly ten years of work, the great building, looking like an enormous half section of pipe, was inched into place in November 2016: it had been constructed 300 metres from the reactor to avoid exposing workers to any more radiation than necessary. The consortium that built it included French and Italian firms, and most of the tons and tons of cement and steel it used were made in Italy and shipped to the site. In all, 28 ships and 2,500 trucks were used to get the components to Chernobyl.[42] That the job was done is a tribute to post–Cold War international cooperation.

Mikhail Gorbachev, general secretary of the Communist Party in the USSR at the time of the explosion, wrote twenty years later that Chernobyl was an "historic turning point" and "perhaps the main cause of the Soviet Union's collapse five years later." After Chernobyl, "the system as we knew it became untenable."[43]

But war continues around the world, even though some thinkers such as Steven Pinker contend that we are killing fewer people proportionately than any time in human history.[44] As the twenty-first century advances, what has changed is the location of the primary battlefields.

If the first 150 years of concrete's use in wartime can be divided into two phases—to counter the effect of exploding shells on fortifications and to protect against nuclear strikes—the turn of the twenty-first century has brought a new phase: concrete's use in urban warfare, both to stop guerrilla attackers and as a background for battles among the ruins of cities.

Indeed, concrete is the most effective weapon on the modern battlefield, according to Major John Spencer, a scholar with the Modern War Institute at the U.S. Military Academy at West Point, New York. "No other weapon or technology has done more to contribute to achieving strategic goals of providing security, protecting populations, establishing stability and eliminating terrorist threats," he wrote in 2016. Concrete may not be "sexy," he says, but given the great possibility that modern warfare will be waged in urban settings, military planners should be asking many questions about how best to use it.[45]

For someone who's interested in the way cities have grown, this is a particularly interesting comment. In the nineteenth century what would today be called city planners or urbanists reconfigured several

European cities, chief among them Paris. Their reasons were several. Relieving congestion that had developed with the arrival of railroads was one, but another was to make military operations in cities more straightforward. Troops could march down grand boulevards far more easily than through the hodgepodge of streets that had grown up over the centuries.[46] The former advantage is no longer a "plus" today, however. In an era when bomb-carrying vehicles and enemies who are just part of the civilian population are features in all combats, wide streets are not a general's friend. Now concrete provides barricades and shelter for fighting on the ground.

Yes, the contemporary battlefield itself is frequently a concrete landscape. A look at any newsfeed or newspaper on any given day shows images of a city under attack. The format of this morning's photo is wide: whoever laid out the page wanted it to be inescapable. In it a man pulling a shopping cart is walking across an urban scene, and behind him are a dozen buildings, blasted apart. Columns on the one on the right are still supporting floors, which must be made of well-placed reinforced concrete. There are no windows left, no interior walls. The only colour is the man's cart, a touch of green in a landscape as gray as the concrete that was poured to build the apartments, offices, whatever, which once promised a better life for their inhabitants.

CONCRETE AND THE HEARTH GODS

I didn't say in which city the decimated landscape I mention above is located because there are so many of them that I didn't want to single out one urban battleground in particular. But even the ruins point, perversely, to another, much more positive relation between concrete and fire, the kind of fire guarded by the hearth gods, no matter what names they are given in your theology.

The apartment buildings of Pripyat where the workers at the Chernobyl nuclear reactor lived are good examples of what I'm talking about. While there are guided tours to some parts of the Exclusion Zone and workers have put in carefully controlled shifts decommissioning the reactors and building the new cover for the radioactive ruins, today the apartments are empty, falling apart, open to the elements, strangers to laughter and love. They have no regular visitors except a few foolhardy "stalkers," generally young people who clandestinely cross into the Exclusion Zone as a challenge, a lark, or maybe even an immunization against fear of an apocalypse.[47]

But when the apartment blocks were built, they were part of a worldwide movement to provide good homes for great masses of people. In the 100 years between the beginning of the widespread use of reinforced concrete for housing and today, several billion people have been housed in buildings that would be inconceivable but for concrete, located on roads that would be equally impossible to imagine without the wonderful stuff.

A caveat is necessary here: the history of buildings seems to concentrate on those designed by architects, which are mostly public buildings or residences of note. A few architects or engineers responsible for ground breaking industrial buildings or the first uses of techniques also get some press. But who knows the names of the men (and for a long time they were only men) who drew up the plans for the buildings that followed? Few do, and practically no one outside their families are aware of the constant work by individuals to improve their lot by building their own dwellings with concrete. More about that later, but first let's try to understand how the buildings like those in Pripyat got that way.

Initially, concrete was viewed as a material that would cut building costs because, it seemed, much less skill was required of the workmen who used it. Someone pouring concrete didn't have to go through the multi-year apprenticeship that was standard for a mason or a stonecutter. With much less training an ordinary guy could turn into an adequate concrete man. (Later architectural historians would comment that the saving in manpower costs on site were shifted to an increase in the amount of work that needed to be done at the design phase, but that did not show up until later.)

The simplicity of the work required is what allowed concrete to become the material of the twentieth century for the suburban do-it-yourselfer like my father, or for the family man in the Third World trying to put together a decent space for living.

Long after I'd grown up and left behind the house my dad had put so much work into (with a lot of help from my mother, I should add),

I spent an afternoon wandering around one of the outlying commuter towns near São Paulo, Brazil. It was at the end of the railroad line that went east from the centre of the megacity. As the train rolled along, I saw small billboards advertising lots for sale, placed so they could be seen both from the train and from the road that ran alongside it. Small houses climbed the hillsides, looking rather like the subdivisions I remembered from my Southern California youth, only in miniature. Not that the people were any smaller, but their aspirations when it came to homes were: square boxes built, it appeared, from concrete blocks or reinforced concrete, many with the rebar still sticking up.

When we got to the end of the line and I got out to look around, I found that the first block of the street leading from the train station had two small shops with sacks of cement and sand in front. The building material was all ready to be taken away and used the following weekend when homeowners had the time to work on improving their little houses. Their constructions were light years removed from the gorgeous reinforced concrete constructions that made Brazilian architecture so acclaimed in the middle of the twentieth century, but they speak to the great need that concrete filled in the lives of Everyman, and Everywoman.

There are dozens of do-it-yourself videos on YouTube about how to mix concrete: a Mexican one showing a pretty young woman shovelling sand, cement, and water together on the ground has been viewed nearly a million times.[48] The popularity of laying down a concrete floor in Mexico for even the most modest houses is linked to government programs, begun in the 2000s, that pay for about US$150 worth of supplies. Now administered by state and municipal governments and far from being universal, the programs have nevertheless produced immense returns in terms of public health improvement. The concrete floors—*pisos firmes*—replaced dirt ones where parasites could breed and be transmitted to the residents, sapping their strength. Children were particularly at risk, and studies have shown a considerable uptick in the school performance of children living in houses with *pisos firmes*.[49]

In Brazil—until recently at least—tens of companies offer quick construction of small houses with reinforced concrete walls. The systems call for levelling a building space, pouring a slab, mounting metal formwork into which rebar is inserted before the concrete is poured on site. The advertising videos show a handful of men installing the forms in a couple of days, with concrete poured and cured within another few.

Then there are modular constructions, where precast walls are delivered to the site all ready to be put into place. The system works best when there is a fair amount of land to build upon, so that houses can be spread out, each with a little space around it for a garden or trees. In other words, it's a system that is going to do little to halt urban sprawl since it recreates on a smaller scale the suburban housing that has overtaken much of the land around North American and European cities. But sprawl usually isn't high on the list of the residents' concerns. Having a new, clean place ready to live in after a short construction period is a big attraction.

That's the same hope that plays into a recent surge of interest by NGOs and some governments in 3D concrete printing. A number of firms claim to be able to build small houses in a couple of days by having 3D printing machines—now used for many other tasks—squirt out concrete slurry through movable nozzles to the rhythm of a computer program. The process looks a bit like a robotized cake-decorating machine, but its promoters say they can build houses for between $1,000 and $4,000.[50]

A Chinese firm, WinSun, has developed more experience with using 3D to build housing: recently it produced several traditional Chinese courtyard houses outside Shanghai using materials recycled from the demolition of large buildings and mine tailings. One observer called the result "painfully ugly," but the company says the technique can be adapted to any style. Proof of this lies in seventeen 3D-printed office buildings that were made in China and then shipped in containers to Dubai, according to WinSun founder and chairman Ma Yihe. He also told a Singapore newspaper, only half-jokingly, in 2017 that his company could build the wall between the United States and Mexico that President Donald Trump lusts for "much cheaper and in less than a year. We could definitely do it. Maybe at around 60 percent of the projected cost and three to four times faster."[51]

One of Ma's motives in developing the technique was to provide low-cost housing in developing countries, he has said. In that he was following other visionaries. The desire to provide sanitary housing for ordinary working families has been the motor for concrete construction since the middle of the nineteenth century, when immense numbers of people attracted by the jobs created by the Industrial Revolution crowded together in cities. Usually what was offered them were lodgings that existed before the great influx began, or hastily thrown together buildings with little or no services. But the migrants to cities were not without a few friends in high places. Among champions of better housing for workers was French emperor Napoléon III, who gave Baron Georges Étienne Haussmann a mandate to carry out a sweeping redevelopment plan. Among the things that Haussmann did was to raze buildings dating from the Middle Ages in order to make way for broad boulevards and more harmonious, hygienic buildings. The changes have been dubbed Haussmannian, but without the emperor's enthusiastic support Haussmann would have been limited to the continuation of projects already begun, plus a little tinkering with traffic patterns.

The emperor had plans to change Paris long before he was elected prince-president of France's First Republic in December 1848. (He became emperor in 1851, but that's another story.) Educated in Germany and Switzerland, he had spent most of his early life in exile, and in the early 1840s published a theoretical work, *L'Extinction de pauperisme*, that drew on his observations of several countries, including Great Britain. His interest in constructing *cités ouvrières* was far from superficial once he had assumed imperial power. When early versions of this kind of workers' housing were criticized by residents, he worked with them to produce a lower, two-storey version for the Universal Exposition of 1867: it was built with reinforced concrete and showcased both the material and French engineering skill.[52]

That year also saw the first completely concrete houses constructed in Winona, Minnesota, although a few builders had been experimenting with the material for several years. One put up a number of concrete buildings in the little town of Seguin, Texas, in the 1850s. He was fascinated by natural cement and like Ward in New York State, one of his motivations was also to build a fireproof house. His constructions had little or no influence on the development of concrete construction, though, because his houses were forgotten for decades since they were covered with stucco, like many adobe brick buildings in the "Spanish" style popular in the region.[53]

The real housing changes that the material facilitated didn't come for a good fifty years, when concrete came into its own after the turn of the twentieth century. Thomas Edison, who was perhaps even more fascinated by concrete than he was with electricity, developed a mould in 1906 that could essentially create a three-storey house in one pour of concrete. U.S. Steel decided the process could hold the key to solving a housing shortage in Gary, Indiana, and used it to build houses for the company's managerial staff. The price for each was touted to be about $1,200 at the time, compared to $30,000 for a similar house made of stone. But start-up costs proved prohibitive, although seventy of the eighty-six houses constructed remain, some still in good condition.[54]

An example of the slow acceptance of concrete as a building material can be seen on the street in Montreal where I live now. Laid out sometime in the late nineteenth century, the first houses on it were built in 1896. Large and elaborate, they are elegant Victorian abodes that must have been surrounded by fields and trees. Significantly, their foundations are made of stone and brick. Around 1910, however, development of the city pushed northward, and with it builders who constructed more modest houses that nevertheless featured what was then the latest in technology. When one family remodelled a few years ago, they discovered that their house, built in 1909, had both pipes for old-fashioned gas lighting as well as wires for new-fangled electricity. In contrast, our house, built three years later and ten doors north, has only electrical wiring, which apparently by then had become the amenity of choice. It also has a poured concrete foundation, which appears to have become the norm by then: it's dug deep below the frost line with the basement floor covered with a not-too-level concrete slab. Planks used as forms for pouring the concrete were reused later for the subflooring, we recently discovered.

When it was built in 1947, concrete footings formed the foundation of the house that my parents lavished so much work on. It didn't have a basement because the climate was so mild that there was no need to dig below a frost line. There was an irony there not lost on the adults around me: just about the time that basements might come in handy as shelters, houses aren't built with them, one of the neighbours said frequently.

But note the date: 1947. It was after World War II, and people were thinking of peace and prosperity, not more war. Concrete construction for the masses was coming into its heyday. The housing shortage was acute because little housing had been built during the Great Depression of the 1930s, and the war stopped almost all civilian construction in the battling countries. Catching up to demand involved two sorts of housing solutions, one more dependent on concrete than the other, but both nevertheless literally underpinned by the versatile material. What is more, they also promised rapid construction.

In North America, the major thrust was for individual houses. Government programs in both the United States and Canada provided low-interest guaranteed mortgages, in some cases without a down payment. The result was a great construction boom with suburban houses spreading like a flood across the landscape, encircling cities. While in large part the houses were framed with wood, they never could have been built as quickly or as well without poured concrete foundations. Nor could they have been accessed without a network of concrete-based roads and streets.

The footprint of these new developments was much larger than it would have been if earlier, denser styles of building had been maintained. But the lending programs required lots of a certain size for reasons that went far beyond a desire to see that each house had a little green space (a sentiment that we'll come across later on). Poor people, particularly poor people of colour, frequently couldn't qualify for loans to buy the more expensive houses built on larger lots. Even more disturbing is the fact that thousands were flatly turned down for mortgages and as buyers for racist reasons, so a basic framework of residential segregation was built into the housing boom. Many cities also cut the roads needed to serve the new, largely white developments through many middle-class black neighbourhoods that had succeeded in being developed. Often the new freeways amounted to barriers designed to physically separate white and black neighbourhoods.[55]

Since roads are so important to the world that concrete has built, a word must be said here about how roads themselves are built. The Romans, whose concrete was so long lasting and spectacular, did not ordinarily use the material for their roads. Their method—at least for the roads that have been uncovered or remain today—was to aim for a straight line and to begin by digging two parallel trenches down what would be edges of the road. Then the stones, gravel, and excavated dirt would be piled up as high as 1.8 metres in the space between the trenches. Minor roads had less of this *agger*, as it was called, and the major roads had more, but in both cases a shallow depression was dug down the middle about 2.5 to 3 metres wide. Its sides were lined with curbstones, and layers of graduated stones and pebbles were dumped in to fill the remaining space. In early days a layer of sand topped the roadbed, but later cement was sometimes mixed with the rubble to make a harder foundation. Then the roadway might be surfaced with large flat stones.[56]

Today, new main road construction starts with clearing the roadbed, compacting the soil, and laying a base of gravel and sand that is a half-metre thick at minimum. Then the road is paved with asphalt or concrete, with reinforcing steel frequently added. Concrete is more expensive, but usually lasts longer. The first concrete road in North America was laid in 1893 in what is now downtown Bellefontaine, Ohio, and is still in good repair, while a stretch of concrete pavement laid 1872 in Edinburgh, Scotland, continues to be used.[57] Paving six inches deep is the usual standard for subdivision roads today, but no one has any expectation that the surface will last anywhere as long as the Roman roads have. Thirty years is considered a good life span for concrete, although careful maintenance can extend that, including adding a layer of asphalt which can keep road de-icing salt from wreaking havoc on the concrete in a cold climate.[58] (We'll consider that particular problem in more depth in chapters 4, "Water," and 5, "Air.")

At the beginning of the twenty-first century the United States had 4.3 million kilometres of paved roads, more than twice as many as it had in 1950, making it the most extensive paved road system in the world. Canada, with a population about a tenth as large, has slightly less than a tenth as many paved roads, or 415,600 kilometres.[59] That adds up to a lot of concrete. The implications of this road building—the fallout, if you like—go farther than that, though. That is because when you build a major road, you literally open the way for more settlements that depend on cars and trucks for the transport of people and goods. Greenhouse gases from transportation account for about 30 percent of all the CO_2 emissions in North America today, and a significant proportion in the rest of the world.[60] That's a problem we'll address more fully in chapter 5, "Air."

One of the emblems of this stunning suburban population growth was the Levittown development near New York City. Started in 1947 and using almost assembly-line construction methods, the Levitt firm built more than 17,000 houses in four years.[61] One of their most innovative techniques was the universal use of slab-on-grade foundations rather than digging basements, which takes much more time. Still widely used, slab construction requires some careful work in preparing the site, such as making sure it's completely level, but that can be done with earth-moving machines. Anyone who has watched a subdivision go up has seen the ballet of bulldozers first removing whatever trees, bushes, and other vegetation are growing on the site, then moving the dirt around until the pad on which to pour the concrete is in place. The technique has much in common with the *piso firme* that has had such an impact on health in poor areas of Mexico, only instead of pouring the slab inside the house, it is the first thing done in construction after the site is prepared.

Thousands upon thousands of houses were built in the decades after World War II using this construction technique. The houses frequently weren't large: the Levittown houses only had two bedrooms, although the lots—eighteen by thirty metres—left plenty of room for expansion. The house I grew up in was about the same size initially: when we bought it from the original owners in 1951, it had two bedrooms, one bath, a kitchen, a living room with what the realtor called a "dining alcove," a garage, and a slab patio that the previous family had covered with a sloping roof. Among the many things my parents did was enclose the patio to make a "family room" that had a door leading to the garage. The street outside until then had only been oiled and when, after a year or so, the city put in proper sidewalks and regular paving, the sloppiness with which the house had been built became evident. The garage was constructed too close to the property line, and the difference between the street grade and the building was a good six inches, with no room to install a gradual driveway. As might be imagined, there was considerable *Sturm und Drang* around the house for a while, but in the end my parents decided to make the best of it by converting the garage into a bedroom and laundry room. Then came the concrete project I remember the most, the laying of a new slab for a bigger patio. A couple of years later my parents hired a contractor to build another bedroom and bath using slab-on-grade construction. So, like many of the houses in Levittown and elsewhere in North America, our house grew, thanks to concrete.

Individual houses like ours were not the development of choice in Europe, although some thought was given to it in several countries. In France, for example, the government underwrote an experimental village in Merlan, northeast of Paris, where examples of several kinds of modest, single-family houses were constructed with the help of Canada, the United Kingdom, the United States, Switzerland, Finland, and Sweden.[62] In the United Kingdom several post-war garden cities outside London were built that had single-family houses as their most frequent form, and some observers cast the debate about which kind of housing to build as political. More conservative forces in society championed low-rise and cottage development, while left wing planners and architects argued for high-rises as more egalitarian, more exciting, more bang for your buck. The only agreement was on the need for new housing to make up for the almost four million housing units destroyed during World War II, not to mention the previously existing housing deficit caused by the lack of construction during the Great Depression.

This is where the vision of Swiss architect Le Corbusier (born Charles-Édouard Jeanneret) comes in.[63] As early as the 1920s he called for an innovative way of constructing cities which separated pedestrian and motor traffic, housing and industrial uses. A widely reproduced drawing from this period showed twenty-storey apartment blocks set in acres of green space, or what came to be known as "towers

Figure 3.2: The Swiss architect Le Corbusier proposed high-rise housing set in parks. His ideas have been copied around the world, sometimes with disastrous results. From Le Corbusier, *The City of To-morrow and Its Planning* (1929).

in the park." His *Plan Voisin*, formulated in 1925, would have seen the centre of Paris pulled down and replaced by towers. Part of the motivation for doing so was to remove what was then called insalubrious housing, epitomized by four îlots of overcrowded, poorly maintained Hausmannian blocks in the centre of Paris. In 1906 the death rate in them from tuberculosis—which at the time was taken as an indicator of extreme poverty—was twice as high as elsewhere. Replacing them would almost magically change the lives of the people living in them, it was argued by many reformers. Le Corbusier himself argued that the city should be remade along more rational lines, which he contended were not only healthier but also more civilized.[64]

Significantly, the *Plan* was never undertaken, and the nearest that Le Corbusier himself got to building high-rise housing was the twelve-storey Unité d'Habitation in Marseilles. Begun in 1947 and opened in

1952, it was aimed at a more or less middle-class clientele. Its 337 apartments were served by a row of shops, a nursery, a gym, and a hotel, with common areas on a floor about halfway up the building. Supposedly Le Corbusier based his plans on a study of standardized human dimensions, with the idea that no matter how large the building, the human scale would prevail.[65]

Le Corbusier also said that his thinking was influenced by a project in Moscow, the Narkomfin building, which was constructed just as he was formulating his architectural ideas in the 1920s. Built of reinforced concrete, the building was intended to house employees of the Commisariat of Finance (Narodnyo Kommissariat Finansov, or Narkomfin in popular parlance) and to display the very latest in architectural design and communal living. Nearly all its two-level apartments were very small and without a kitchen: the idea was that cooking, socializing, laundry, and childcare would be done in shared spaces, not necessarily because of space limitations but to get women out of their traditional roles. The Guardian's 2015 series on the fifty most influential buildings in the world rated it twenty-ninth, quoting Lenin: "Petty housework crushes, strangles and degrades . . . chains her to the kitchen . . . . The real emancipation of women, real communism, will begin only where and when an all-out struggle begins . . . against this petty housekeeping."[66]

When Stalin came to power shortly after Narkomfin's completion in 1932, the social experiment was stopped as too "leftist": no more examples were built and those who were lucky enough to live there began changing the apartments into more conventional units, adding makeshift kitchens. The building had been built on stilts, called pilotis, which Le Corbusier used later in his Unité d'Habitation in Marseilles, and which were incorporated into residential housing in Brasília, the Brazilian capital, by disciples of Le Corbusier. This open first level was filled in with shops early on in Narkomfin's history, however.[67] Despite the changes, the building still exists, and in 2018 it began its newest incarnation as a luxury apartment house. The old communists who lived there at the beginning would be shocked, probably: the cheapest apartments were then selling for three million rubles, or US$486,000.[68]

Ideas in architecture and city planning, as we'll see, sometimes are passed around in surprising ways. While Le Corbusier's fame began to grow after World War II, his ideas about housing had inspired social democrats in Sweden well before the war. Beginning in 1932 at a time when Sweden's housing was considered among the worst in Europe, the social democratic government of the day wanted to create a more egalitarian society. It held a competition for urban design in which Le Corbusier proposed a plan to raze much of Stockholm, just as he had advocated major surgery to Paris in the *Plan Voisin*. Although the Swedes rejected his plan, his idea of high-rise blocks of flats found their way into many subsequent development projects.[69] Among the most evident are the housing towers built in the Danviksklippan suburb of Stockholm and opened in 1943. They were featured in a special issue of the British publication *The Architectural Review*, entitled "Sweden: An Outpost of Peace in War." After the war, architect Gordon Logie wrote admiringly: "While other nations have been talking about building landscapes of towers, the Swedes have quietly gone ahead and done it."[70]

The British followed in the post-war period, as did the French when they constructed HLM (moderate-rent housing) on the periphery of cities like Paris and Marseilles. Notably, only one truly tall tower was built inside central Paris: the fifty-two-storey Tour Montparnasse completed in 1973, around which swirled so much controversy that the city government at one point considered its destruction. (La Défense, the high-rise complex outside the western boundaries of the city, will be considered in chapter 5, "Air.")

Housing construction stopped in the Soviet Union during the war, as it did in other countries involved the fray. (Sweden, it should be noted, was neutral during the war.) But afterwards the USSR also plunged in, aiming to provide housing for its population as quickly and as cheaply as possible. Most of the construction initially was brick and board, but in 1954 Nikita Khrushchev, then first secretary of the Communist Party Central Committee, came out forcefully in favour of switching to prefabricated concrete construction.

A liking for concrete and cement was nothing new in the Soviet Union. F.V. Gladkov's novel *Cement*, first published in 1925 and considered a classic of Soviet literature, portrayed the material as a symbol of industrial progress. Gladkov paints the reopening of a cement factory, closed down during the turmoil of the revolution, in exalted terms. "The myriad crowd yelled and thundered . . . . They were dancing and leaping there beneath the high platform, on the rocks and mountain slopes, where the banners flashed liked wings of fire, and the bands rang like thousands of grand bells."[71]

Be that as it may, Khrushchev's speech, Adrian Forty notes, was probably the only time ever that a leader of a country waxed eloquent for five hours about concrete.[72]

Post-war Soviet housing was modest: the model was a five-storey apartment block without elevators, frequently put together in a slapdash way. Forty says that the buildings were nicknamed *krushcheby*, a play on the Russian word for slum, *trushchoby*, and Krushchev, but in more recent years Krushkevka has apparently become a sometimes affectionate name for any five- or six-storey apartment building. The Soviets changed the style of housing in the late 1950s, when it was found that the savings in buildings without an elevator were offset by other expenses. Thereafter the main thrust of the housing program was to construct sixteen-storey buildings with lifts. Between 1955 and 1964 a quarter of the Soviet population, or fifty-four million people, were rehoused in the new flats.[73] At the height of the construction program, some thirteen million people were employed building the apartment blocks, and work continued even after the end of the Soviet Union in 1990.

Some of this housing was comfortable and well appointed. Certainly that built at Pripyat for the workers of the Chernobyl nuclear power facility was. Thirty years after the explosion, the derelict buildings where the reactor's personnel lived show considerable care in design and what are now called amenities: large windows, kitchens with what must have been top of the line equipment, sports clubs with swimming pools, shops. A series of photos from before the nuclear disaster show clean streets, pretty women on a municipal beach, kids in a playground, and dozens of ten-storey blocks of flats that—from the exterior at least—look a lot like better housing developments built during the mid-twentieth century around the world.[74]

As for construction elsewhere in the Soviet Union, people who now live in the post-war apartment blocks are sometimes adamant in their attachment. In 2017 when the Moscow municipal government, with the support of President Vladimir Putin, announced that it would pull down 7,900 of the apartment blocks to replace them with flats that were 20 percent larger, many residents were not pleased. Some, who'd moved in when the flats were newer, said they loved the small apartments, which were originally planned with a thirty-year life span. The flats had been a giant step forward for many who had lived in shared quarters for years as they waited for their names to come to the top of housing waiting list. They were as delighted as were the first residents of the United Kingdom's early high-rise towers where, as one woman told John Grindrod, "Does tha know love, we thoughts we'd died and gone to heaven . . . . Toilet were i' there, not at t'bottom of yard."[75] Now many of the Muscovites fear that their neighbourhoods will be razed without concern for the street life and social connections that have grown up over the years. Even the prospect of extensive repairs—the concrete to make walls was poured around heating and water pipes in some buildings, and replacing them would be an immense task—doesn't faze them.[76]

The American urbanist Jane Jacobs would understand that attachment to established neighbourhoods. When she published her *The Death and Life of Great American Cities* in 1961, the United States was engaged in a spasm of pulling down old substandard housing and replacing it with Le Corbusier–inspired apartment high-rises. The process was tainted in several cities by the desire to keep internal migrants, who frequently were African Americans, out of formerly all-white neighbourhoods. The result was the death of old, often vibrant neighbourhoods and the construction of high-rise ghettos.

Jacobs pointed out that a mix of old and new buildings allows for people of different incomes to live together, and for businesses to start up with relatively little capital. "Large swatches of construction built at one time are inherently inefficient for sheltering wide ranges of cultural, population and business diversity," she wrote[77]. Even corner grocery stores have a hard time making a go of it if they must pay the kind of rents that owners of new buildings expect. That's the reason why the space sometimes put aside for small stores in high-rise public housing is rarely used for convenience stores: the volume of business isn't large enough to cover the costs. A certain level of density and mixed uses is extremely important, Jacobs argued at a time when many of the big guns in North American urban planning were championing low-density suburban development where the ideal was a development of individual houses on lots big enough to have substantial gardens.[78]

Her admonitions went unheeded by decision makers for the better part of two decades, during which time many of her predictions came true as several public housing projects became unlivable and were eventually torn down. Not only were some of the problems inherent in Le Corbusier–style blocks—lack of "eyes on the street" surveillance of the building's corridors by residents, for example—housing authorities decided not to keep up maintenance in many of the projects. Not having janitors to clean common spaces or to repair broken elevators invited vandalism and eventually crime. In addition, as one British critic rather prissily (but accurately) noted, "the areas where adults expect to be responsible for their children's behavior have not been firmly established" in either buildings or in the development at large.[79]

The most infamous of these developments were the Igoe-Pruitt houses in St. Louis, Missouri, where initial hope gave way to terrible conditions within a few years' time.[80] The development had become a dumping ground for the poor, who were largely African American during a period when the white population of cities all over the United States was fleeing to the suburbs. So damaged and hated did the complex become that it was razed in 1973, not twenty years after it was built. Similar design flaws and damage to the social fabric in the United Kingdom led to demolition of some of the early projects at about the same time too.

Then came a major political change when conservative Prime Minister Margaret Thatcher launched a movement to get rid of public housing in the United Kingdom through the Right to Buy scheme. Beginning in 1980, residents of many council flats could buy their apartments at discounted rates, neatly getting the government off the hook for repairs and renovations. At the same time, almost all new investment in public housing was stopped. The result was that the publicly owned rental flats became increasingly where the poorest people lived.

This is not the fault of concrete, of course. It simply is the material used to build these unloved buildings. A few exceptions to public housing programs exist, notably the one in Vienna which dates from the 1920s but into which much money and concern has been poured over the decades. Currently about 62 percent of Vienna's residents live in social housing where rents are regulated and anyone making up to the equivalent of US$53,225 after taxes is eligible for a subsidized apartment.[81] The reason why the housing works has everything to do with continued government support for it, and little to do with what it is built from.

To see what kind of housing concrete can provide when serious attention is given to how people would like to live, one only has to visit Habitat 67 in Montreal.

On an early summer day as brilliantly clear as the one when I visited McInnis Cement I took a guided tour of Habitat, a model housing development built as part of Montreal's International Exposition of 1967, Man and His World. It wasn't the first time I'd been there: two of my husband's colleagues had rented apartments there in the early 1970s and we'd been invited to a couple of parties. I must admit I was taken with the building very much at the time and we talked about trying to see if there was anything available next time the lease came up on our downtown place.

This time I was one of four architecture-lovers who were shown around the complex by Judith Bradette, an art historian who is in love with 1960s architecture and the exuberance of Montreal during the period. The tours were begun in 2017 as part of the celebrations of the fiftieth anniversary of Expo 67 and give people like me who've always wanted to see more of the place a great chance to do so.

Set on a small spit of land that separates the harbour of Montreal from the St. Lawrence River, the complex was designed by a young man who wanted to combine the best of dense city living with the pleasures of suburban greenery. In the early 1960s Moshe Safdie had received a fellowship to travel around North America looking at housing. Decades later he told an interviewer from *Architecture Daily* that what he found was "public housing that nobody wanted to live in, that were centers of violence and unhappiness, at the same time that Le Corbusier was designing l'Unité d'habitation in Marseille, that seemed to betray all the values that he had espoused in the 1920s and 1930s, a big complex building with none of the qualities of house or village."[82]

This was not what people wanted or needed, Safdie concluded, and when he came back to complete his architectural studies at McGill University, his idea was that by designing a basic concrete module that could be produced in a factory, he could build relatively low-cost housing that could be assembled to suit the needs of many sorts of households. Each unit would have at least one garden space, and units that combined more than one module would have more garden too: his motto was "for everyone a garden." The modules would be placed one on top of another, and connected by walkways that would be like the streets of a small town or cohesive neighbourhood. He developed his first model as his master's thesis at McGill University, and then, at age twenty-five while working to oversee the world's fair master plan, he was able to build a prototype. It became one of the most visited of the fair's attractions and was selected as the heritage element to be saved after the fair was over and most of the other pavilions were torn down. In being chosen for this honour, Habitat 67 joined other emblematic buildings left over from world fairs, such as the Space Needle in Seattle and the Eiffel Tower in Paris.

The modules were each about 183 square metres in area and measured 11.8 by 5 metres with a height of 3 metres, which worked out to 2.4-metre ceilings: all the wiring, pipes, and heating and cooling ducts were hidden by a false floor covered with light wood parquet. The modules were produced at a concrete batch plant on site, where concrete was poured into one of three moulds. When the module had cured it was transported by a crane on rails: each weighed between 6,500 and 8,500 tons and was hoisted into place in a carefully calculated progression.

Safdie thought a single module unit would be suitable for one person or a couple: initially the apartments included units of up to four modules designed for larger families or for people who wanted more room. One five-unit module has been arranged since then, as people have expanded their space by buying adjacent modules. Safdie's own four-module apartment is on the top floor and, restored to the way it was in 1967, is part of the guided tour. It, like the whole of Habitat 67, was classified as an historical monument by the Quebec government in 2009.

Sleek, clean, and inviting were the words that came to mind when I entered Safdie's *pied-à-terre*, which is now free of furniture to show the apartment at its essential best. On two levels, it has four bedrooms and two baths on the upper one, and downstairs, a living room, powder room, dining room, kitchen, and a family room/den, plus four terraces, each one about twice the size of a balcony on an ordinary high-rise building. Windows pointed in all cardinal directions so light flooded into the apartment all day long. Down below on the east side, the St. Lawrence rushed past. Across the river the green trees of the South Shore suburbs were visible while three of the Monteregian hills stood high above the flat land. To the west the skyscrapers of downtown Montreal were silhouetted against Mount Royal. Roses bloomed on the terraces of the adjoining apartments, bright annual flowers had just been set out in the built-in flower boxes, while swallows swooped to catch mayflies hatching on the river and red winged blackbirds called to one another.

Judith emphasized how much Safdie hoped his complex would become a real community and a model that could be reproduced around the world. According to its official description, it was supposed to operate as a "sector of a city in which residential, commercial, and institutional facilities are integrated within a single complex . . . . Large inclined surfaces, like grand hillsides, form the complex's residential component, oriented toward unobstructed views and sunlight. Partially sheltered on the ground below are parking and transportation facilities, shops, schools, office space, and an extensive network of parks that connect with the surrounding city."[83]

The idea of stackable modules sounds a bit like a Lego ad, and in fact Safdie says that he and his team bought up "all the Lego in Montreal" to work on the plan for Habitat.[84] But the way the modules are put together is much more sophisticated than a block set: all parts of the building, including the modules themselves, the pedestrian

Figure 3.3: Habitat 67 at fifty: Moshe Safdie's groundbreaking attempt to create housing that included "for everyone a garden." Photo: Mary Soderstrom.

walkways, and the elevator cores, are load-bearing parts of the building. The units are connected with high tensile rods, cables, and welding, so that some modules are actually suspended from ones above.

Safdie proposed building one 12-storey section and one 22-storey section with a total of 1,200 housing units, a 350-room hotel, 2 schools and a neighbourhood shopping centre. In the end, however, only 158 housing units in a 12-storey section (including two subterranean floors for parking and technical needs) were built.[85] And while he hoped that the idea could be successfully transplanted to much less privileged contexts, including projects for modestly priced housing in New York, Puerto Rico, and Israel, no other Habitat was ever built.

Back when we had been charmed by our friends' apartments, rents were hefty but not beyond the reach of a young couple with a decent middle-class income. But Habitat 67 was losing money at that point, and in the 1980s the Canada Mortgage and Housing Agency, the federal government agency which had been underwriting the project, bailed out, letting a non-profit corporation take over. Now residents own it in undivided co-ownership, with their service fees (including property taxes, heating, and cooling) based on the number of modules in their apartment. In 2019, the fees for one four-module unit that was for sale came to c$41,988.00 annually, and the owners were asking c$1,390,000.[86]

That's a whole lot more than most folks starting out today could afford without serious help from family or phenomenal success with a high tech start-up, which explains why the average age of the people we saw during our tour must have been in their sixties. The only exceptions were maintenance staff, who seemed to be very busy cleaning the public spaces and doing things such as repairing small cracks in a ground floor concrete walkway. Obviously keeping a massive concrete structure in good repair takes a lot of work and is expensive. That was brought strikingly home as we looked upriver from Safdie's living room aerie. Work was proceeding, we could see, on the new Samuel de Champlain Bridge, which we'll talk about more in the chapter on water. Suffice to say for the moment that the old Champlain Bridge, made of prestressed concrete and steel, was opened five years before Habitat, but has weathered so badly that it had to be replaced for safety reasons because it was literally falling apart.

Figure 3.4: The view from Moshe Safdie's terrace in Habitat 67, with the Victoria and Champlain bridges in the background. The former was still solid after 158 years, but the latter, at only fifty-six years, was being replaced. Photo: Mary Soderstrom.

On the other hand, Habitat 67, while it has needed cosmetic repairs, remains structurally sound and is being regularly tested for strength and possible problems, our guide stressed. The concrete surfaces show repair lines here and there, and wear and tear at some particularly exposed corners, but the building remains the pleasing, textured mass that it was fifty years ago, contrasting beautifully with blue of sky and water, and green of the gardens and riverside park that surrounds it. Upkeep matters, as does good original design.

But looking back I remember that neither rent nor an aversion to gray masses of concrete kept us from moving to Habitat 67 all those years ago. It was the fact that there was no shopping in the complex, infrequent bus service, and isolation from the city which seemed so close when viewed from the windows. At that point I was becoming a born-again pedestrian, enjoying the ease with which we could get around Montreal without a car. And, while we had no kids at the time, I saw how having small children would mean daily car trips with Mom or Dad to get to school or day care. Nor did I have any idea of the economics of local commerce, but I could see Safdie's creation was far from being a perfect solution to the problem of density and the desire for greenery. In more recent years, as I've learned more about city planning, it's become clear to me that many, many schemes to create model housing for ordinary folk may be great for the people lucky enough to live there, but turn out to be far too expensive for anyone but the very well heeled.

Don't get me wrong, I'm well aware that high-rises—which are now the default pattern for high-density city housing—can be pleasant places to live. For proof, all we have to do is look to those designed for the wealthy like the Trump Tower in New York. The tallest reinforced concrete building in the city when it was built in 2003 (262 metres high), for three years the Trump Tower was the New York headquarters for President Donald Trump and decorated in sumptuous style.[87]

Concrete towers were also built for the middle class in some cities. Between 1960 and 1980 Toronto saw 3,200 apartment blocks of five or more storeys built in its suburbs, perhaps the biggest private experiment with the "tower in the park" model anywhere. I remember being astounded at the forest of high-rise buildings we encountered as we approached the city from the west for the first time in the summer of 1968. The map said we were far from the urban centre, but there they were, carefully planned, self-contained new high-rise communities. I'd seen nothing like it in California, where we came from, and I must admit I liked the buildings' sleek lines. Others thought the same, and scores of the buildings, privately financed, surrounded by green space, far away from the aging core of the city, went up in suburban Toronto. Today, the Greater Toronto Area (GTA) has the highest concentration of high-rise buildings in North America after New York.[88]

Toronto currently is in the middle of a new round of skyscraper construction, but many of the older buildings have not aged well. Most of the older high-rise neighbourhoods have become what has euphemistically called "affordable" housing, particularly for recent immigrants. As time passed, problems developed and new owners frequently didn't correct them. Elevators broke down, swimming pools and tennis courts were abandoned. Now census data says that more than 70 percent of the million or so people who live in the GTA's suburban high-rises live in poverty.[89]

At the moment, North American residential high-rise construction is mostly condo construction, and how it will fare remains to be seen. Continued maintenance, as well as good construction to begin with, will make a difference. A little social engineering might help too.

In a book about concrete, talking about engineering that works on people as opposed to working with materials might seem a little strange. But that appears to be the key to the success of one of the few places (and the only one I've seen up close) where a city of concrete high-rises has worked, still works, and promises to keep on working: the island-nation of Singapore. When it broke away from (or was thrown out of, as some histories have it) the Malay republic in the 1960s, a quarter of the population lived in extremely crowded two-storey buildings called shophouses or in flimsy constructions on the edge of waterways.[90] You can get an idea of what life was like then in a small museum at the headquarters of the agency that rehoused Singapore, the Housing Development Board (HDB). Large photos of shacks built on pilings over water and grubby little rooms with children asleep three to a cot are on display. So is a kitchen from a shophouse, complete with original furniture (neatly tagged with the names of donors) sitting with legs in pans of water to keep insects from climbing up and getting to the food.

But now about 80 percent of Singapore citizens live in well-cared-for, rather spacious HDB flats, of which more than 90 percent are owned by the people who live in them. Take the town of Toa Payoh, which was the first completely planned HDB town and which I have strolled around a couple of times. Certainly it wasn't a promising location back in 1960: the name means "big swamp" in the Hokkein dialect of Chinese, which was then one of several languages spoken in Singapore. Nevertheless, squatters in those times of great housing shortage had already built make-shift houses on it: more than two

Figure 3.5: Toa Payoh is the first town to be fully planned and built by Singapore's Housing and Development Board. Singapore is a rare case of successful use of "tower in the park" urban planning. Photo courtesy of Housing and Development Board, Singapore. Used with permission.

years of work were required to clear the land for construction. Then Phillips, the electronics firm, was encouraged to put up a factory as the first housing was constructed: the prime minister of the day, Lee Kuan Yew, wrote in his memoirs that from the beginning he wanted land "set aside in these estates for clean industries," which would then tap the large pool of young women and housewives whose children were in school.

Just across the street from the Toa Payoh metro station stand the town's first blocks of flats. From the outside the buildings look new because a major upgrading was undertaken here about twenty years ago, while another one, the Remaking Our Heartland program, is now underway.[91] It's that continued upkeep—along with the policy of home ownership and ethnic mixing—that underlies Singapore's success. Today, the town is home to more than 100,000 residents who live in more than 37,000 HDB-managed flats. On a Thursday afternoon—a half-holiday for many school children—the station will be bustling with kids in school uniforms carrying sports equipment and horsing around. The shopping concourse opens up onto a larger commercial area filled with kiosks and more shops. You'll find good local food here in the food court, made on spotless premises: one of Singapore's "clean" initiatives has successfully forced very high hygienic standards on "hawkers"—the small food vendors. There's a busy McDonald's too, but people will also be eating much more interesting and varied cuisine drawn from Singapore's Chinese, Malay, and Indian culinary heritage. The faces in the crowds reflect that heritage because it's HDB policy that each building have residents from each ethnic group. Cynics have said that's to avoid creating voting blocks based on ethnicity, but more importantly it means that there are no ethnic ghettos.

Whatever the underlying reason—and there may be a bit of both involved—the fact is that nearly all Singaporeans own the apartments where they live, which gives them an enormous stake in making the buildings work. The incentives to buy are very strong. The government forces people to save through deductions at source to the Central Provident Fund, and housing is one of the few things besides pension that they're allowed to spend their CPF savings on.[92] (It should be noted that as Singapore grows richer and richer, there is a disconnect developing between the residents of this housing and the minority of people who have become enormously wealthy in the last twenty years. Kevin Kwan's *Crazy Rich Asians* trilogy shows this vividly: one of the major characters grew up in the Toa Payoh neighbourhood and is looked down upon by the snobbish family he marries into. We'll touch upon that when we talk about Moshe Safdie's more recent projects in the chapter 5, "Air.")

Singapore's home ownership scheme had other goals too. Lee Yuan Kew writes in his memoirs that part of its motivation was to give "all parents whose sons would have to do national service a stake in the Singapore their sons had to defend. If the soldier's family did not own their own home, he would soon conclude he would be fighting to protect the properties of the wealthy."[93] And behind the requirement that all young men put in two years of national service lay two other nation-building policies: protecting the small country from its neighbours—being kicked out of Malaysia was a traumatic experience, Lee says—and exposing young people from different backgrounds to the same "Asian values" which are also stressed in school.

Note the difference between this and the British housing acts in the 1980s which allowed inhabitants to buy some units at below-market cost. The United Kingdom did not have a compulsory universal savings plan like Singapore's, which meant that the chance to buy was only open to people who had enough money already saved to meet mortgage requirements. Poorer people were by and large left out, and the blocks where they lived deteriorated because public support for maintenance declined at the same time.

While these negative consequences of "tower in the park" planning have led much of the world to turn away from this model of high-rise social housing, the buildings continue to be constructed in China, and on a scale unimagined elsewhere.

As it happened, not long after I explored the modest streets of builder-made houses outside São Paulo, Brazil, I went to Shanghai. I was working on a book about the complicated way people interact with nature in urban environments and I wanted to use these two remarkably different megacities as examples of the varieties of ways this urge for green plays out. This is not to say that high-rise apartment buildings are absent from the Brazilian city, but they are, with only a few exceptions, the kind of housing that would count as luxury in a country where the beauty of public concrete construction is legendary. (The legacy of Oscar Niemeyer, in particular, is something we'll talk about in chapter 5, "Air.")

Shanghai—and indeed all of China—at the time I visited was well advanced into a movement to rehouse the population so massive that it seems almost impossible. If rebuilding Europe and the Soviet Union after World War II was a colossal undertaking, the Chinese effort is orders of magnitude greater. As Wade Shephard, author of *Ghost Cities of China*, writes, "when the Communists came to power in 1949 the country was in shambles. In fifty years China saw the rapid decline and fall of the Qing dynasty, warlordism, attempts by the Nationalist government to take control, invasion and occupation by Japan, and a civil war that tore the country apart. Mao essentially won the right to rebuild a land in ruins."[94] To do this, Mao's government used many of the methods used by the Soviets in their post–World War II rebuilding: five- and six-storey housing blocks without elevators, built as quickly and as cheaply as possible. Concrete was the material of choice for the same reasons it was used elsewhere: its wide availability, relatively cheaper cost, and ability to be put up with less-skilled labour. Some of the cement was made in small kilns scattered around the country, and quality was frequently lacking.[95] As Shephard comments, "what was built then was not meant to last forever."

But some of the housing dating from that era that I saw in Shanghai in 2005 was still being lived in—and improved—by residents. I spied evidence of this in the new air conditioner units, well-tended courtyard gardens, and fresh paint I saw as I made my way around the city on foot and on public transportation. The rows of even older, traditional two-storey houses clustered around a courtyard still buzzed with life. It was April, and in several *longtangs*, as these small urban enclaves are called,[96] old-timers were carefully tending to tomato and pepper seedlings which probably had only recently been set out. Riding the elevated trains on the Pearl Light Rail line allowed me to look into the windows of families gathering for a meal on the third or fourth floor. I saw smiles, plants in a little pop-out cage attached to a window, more fresh paint.

But the wrecking ball and heavy equipment were waiting in the wings. One evening I spent a couple of hours watching from my hotel window as bulldozers attacked a block of traditional housing across the street. Workers had been at it since about sunrise, and work was continuing under floodlights. About midnight it ended with a great wave of nauseating stench that rose up when the ancient sewage system—underground cisterns, perhaps—was breached. But by sunrise the smell had dissipated, and when I looked out then the site was coming to life as the workers, who apparently had slept on-site in a row of trailer-like barracks, began their day.

Scenes like that have been repeated thousands if not millions of times in Chinese cities since the late 1990s. So have well-orchestrated attacks on farmland on the outskirts of cities, as totally new communities sprang into being. The goal was to rehouse urban dwellers whose lodgings are substandard or falling apart, and to provide urban housing for 300 million migrants from rural areas by 2020. That's like building new homes to house twice the population of Russia, or 95

Figure 3.6: Traditional courtyard housing being demolished to make room for new high-rises in Shanghai, 2005. China's goal is to rehouse urban dwellers whose lodgings are substandard or falling apart, and to provide urban housing for 300 million migrants from rural areas by 2020. Photo: Mary Soderstrom.

percent of all residents of the United States, but as of this writing it looks like China is right on target. One need look no farther than this immense housing boom to find the reason for the skyrocketing consumption of cement and concrete in China.

The usual model for this new housing is overwhelmingly Le Corbusier's famous "tower in the park." Shepard quotes Bianca Bosker, author of *Original Copies*, a book on architectural and design trends in China: "What's fascinating about these places is that China has not opted to copy the latest and greatest in architecture or technology which it easily could have. Instead it's replicated . . . outdated design principles the rest of the world has long soured on. Those gated communities, mostly located outside the city centre, exacerbate China's urban sprawl. And ecologically speaking they're a total disaster: water heavy, land intensive and deeply car dependent, they replicate some of the most problematic urban design practices."[97]

There are a few exceptions to this pattern, and, interestingly, Moshe Safdie is responsible for two of them which draw on the until-now-unrealized potential of his Habitat ideas. One is the Golden Dream Bay project in the coastal city of Qinhuangdao, 322 kilometres east of Beijing. Its port is ice-free and handles an immense amount of coal mined in the mountains to the north, as well as coke, petroleum, and timber. But it also has long been the site of a beachside retreat where Communist Party officials escaped (and still escape) the heat and pollution of Beijing, and the Safdie project is located on the same stretch of coast.[98] The other project is Raffles City in Chongqing, an inland city of thirty million located at the confluence of the Jialing and Yangtze rivers. The two cities are widely separated geographically, but have in common relatively good air: they are ranked 44 and 45 out of 74 big Chinese cities on a scale where number 1 has the most polluted air.[99]

This time the problems of scale which made Habitat 67 uneconomic—not big enough to allow for the development of economies of scale in construction, nor with a population dense enough to make transportation and commerce viable, for example—are being met by going much higher. At the same time, the buildings have been "fractalized"—broken up into small units that are placed so that each has good exposure to light. The buildings are not conventional high-rise towers but usually step back a level at a time so that the finished product looks rather like a hillside or an A-frame.

The attraction of even the most conventional high-rise housing in China is strong. Before the great change in China's economic strategy in the early 1990s, all housing was owned by the state. After 1998 people had the option of buying the houses they lived in or of being bought out and moved to new housing elsewhere in their city. As architect Christopher Choa told me in 2005, a family who had been sharing a kitchen and toilet with eight other families could get the equivalent of US$70,000 for giving up their home. That money could buy a three-bedroom, two-bath apartment in a high-rise, and few were the families who could resist, even if it meant being relocated far from friends and work. A dozen years later, Shepard writes, most of these old central neighbourhoods had been razed and redeveloped, and prices of new housing had skyrocketed. Nevertheless, in 2018 (the latest figures available) 89.6 percent of Chinese owned their own homes (apartments for the most part).[100] That's the second highest percentage in the world: Singapore comes first with 91 percent and Russia is in third place with 87 percent. Canada and the United States are well behind, at 66.5 and 64.2 percent respectively. What is perhaps even more surprising is that many of these people don't live in the homes they own, having bought a second or even third one as investments. Given rising real estate prices, property is perhaps the safest, most lucrative way to store your money. In addition, as in Singapore's mandatory savings scheme that has financed its housing programs over the last half century, both employees and employers contribute to China's Housing Provident Fund, and the money saved there can only be used for housing until a person leaves the workforce, at which time it becomes a retirement nest egg.

Ghost cities are what some observers have called much of this Chinese development. Certainly at nightfall when I was in Shanghai there were many buildings where practically no lights shone, and when I walked in some suburbs under construction I saw few people besides the men and women working on the building sites. Foreigners, including an Al Jazeera news team who have visited new cities like Orodos Kashbankshi in Inner Mongolia, have decried this construction as evidence of monumental miscalculations on the part of the Chinese government. They argue that this series of "ghost cities" are not really doing much to better house ordinary Chinese, in part because the construction is extremely shoddy. The new developments also wreak environmental havoc both by encroaching on agricultural land and by being directly responsible for the immense amounts of CO_2 emitted in producing the concrete required to build them and by the cars and trucks used to serve them.[101]

Shepard, who agrees that there is a lot of poorly built housing in China today, counters the "ghost city" criticism by arguing that this enormous effort by the Chinese to shift millions to new urban centres means that the cities don't grow like cities elsewhere have grown. The housing and office buildings have to be there first. Then come schools and a commercial centre when the buildings begin to fill up. Ordos Kangbashi for example—where aerial photographs in 2010 showed empty streets—has busy sections now and was home to more than 150,000 people in 2018. While that's less than the 300,000 it was originally slated to house, Shepard says that might come later. He writes in an article in *Forbes* magazine:

> What was completely missed in the analysis of this initial report was . . . Kangbashi was a mere five years old. This means that a massive section of an entirely new city was built and partially populated within half a decade. In a world where . . . the Empire State Building was once despairingly monikered the "Empty State Building" . . . and where it regularly takes western cities five to ten years to build civil works projects like monorails or new subway lines, Ordos Kangbashi probably should have impressed the world with its rapid pace of development. Instead, it was mocked as a ghost city.[102]

The idea of building completely new cities is not a new one, even if the Chinese effort is without precedent. The middle of the twentieth century saw two noteworthy efforts that, in addition, demonstrated the versatility of concrete: Chandigarh in India and Brasília in Brazil. In both cases the new cities were designed to be showcase capitals and they are: both Chandigarh and Brasília have been declared UNESCO World Heritage Sites.

In India, Chandigarh rose from the Punjab plain to become the capital of the region that, before Indian independence, had been governed from Lahore in what is now Pakistan. In 1947 the formerly British colony was partitioned along religious lines, with the vast centre of the Indian subcontinent becoming India, a mostly Hindu nation. Pakistan was formed from two large, mostly Muslim regions on the eastern and western extremities of India with Lahore as its capital. The new Indian government saw clearly that what had become the state of Punjab now needed a new capital, and intended that it to be a wonder.

Authorities first turned to American architects Mathew Novicki and Albert Mayer for plans, but when Novicki was killed in an airplane crash and Mayer decided to withdraw, pressure was put on Le Corbusier to accept the challenge. Initially reluctant, in 1950 he agreed when guaranteed that his cousin Pierre Jeanneret could oversee the actual construction of the new city. In the end, Le Corbusier found both the project and the country invigorating, even though, in those days before jet airplanes, travelling there and back from France, where he remained based, was exhausting.

Le Corbusier had strong backing from Indian Prime Minister Jawaharlal Nehru: the pair shared many ideas about what one historian has called "humanitarian socialism."[103] The challenges were great however, and Le Corbusier and his associates worked fourteen years to create an administrative centre along with housing for a half million people of all classes who would work in it.

As he began to draw up plans for Chandigarh, Le Corbusier was finishing up his famed Unité d'habitation in Marseille, which, as noted before, is much lower and smaller than the "towers in the park" that he had frequently previously proposed. While he was proud of this model housing project, what he planned for northern India was much grander.

Laid out on a grid, the city was to include public spaces, gardens, and many street trees. The result six decades later is that Chandigarh today feels much less crowded and hectic than other Indian cities. The heart of the city, the Capitol Complex, contains the Legislative Assembly, Secretariat, and High Court buildings. These are built with carefully worked concrete, a material that at the time was relatively innovative in India, and which was also a relatively expensive material. Hospitals, schools, commercial centres, parks, and housing were planned meticulously and built mostly according to the plan, with the city divided into numbered sectors. Residential sectors were to be served by their own small commercial zone, schools, and other services. Fourteen grades of housing were built, with a "minimum house" for people at the bottom of the socio-economic ladder. The smallest unit was to contain at least two rooms and be 9.3 square metres (100 square feet): by way of comparison, a trendy North American "tiny house" today measures 9.3 to 37.2 square metres (100 to 400 square feet). In Chandigarh, a kitchen, veranda, bathroom, wc, and back and front courtyard were the next steps up the housing ladder.[104] The housing at the bottom of the scale, it should be noted, was to be built with local brick because concrete was "considerably more expensive."[105] That's a far cry from the situation in North America when I was a child or in India or Brazil today, where sacks of ready-mix are easy to come by.

Le Corbusier had no direct role to play in the building of Brasília, the Brazilian capital, but his influence was felt there nevertheless. In 1929 he had spent some time in Rio de Janeiro, then the capital of the country, lecturing on his ideas about cities. He returned eight years later to consult on designs for a new university campus, and a Ministry of Education and Health building. During his visits he gained several converts to his ideas, including Lúcio Costa, who would submit the winning plan for a new Brazilian capital in the centre of the country.[106]

The political reasons for building Brasília were different from those which led to the construction of Chandigarh. The idea of a new capital had been around since shortly after Brazil was settled in the sixteenth century, because it was thought that an invader might attack Rio from the sea. But both projects had national affirmation as common motivators. Building beautiful, modern cities with concrete, then considered the most exciting material around, was clearly an idea that had much appeal. In 1955 presidential candidate Juscelino Kubitschek promised that if elected he would build a new capital city by the end of his five-year term, and shortly after he won he launched an international competition to choose the new city's plan.

What that would look like was to some extent determined in advance, however, since modernist architect Oscar Niemeyer, who designed the United Nations headquarters in New York, had already been chosen to design the major buildings. Costa, the competition's winner, nevertheless was something of a surprise, since his plan, unlike the voluminous, detailed ones submitted by other contestants, was presented on four large cards without a technical plan or drawing attached. He took the relatively flat land of Brazil's central plain and proposed a spread-out city that would be monumental in scale. The central axis would feature broad, automobile-friendly thoroughfares along which the buildings of government would be set, housing its executive, legislative, and judicial arms, as well as all the ministries. Extending out from either side would be curving boulevards of housing, the north and south wings, the Asa Sul and Asa Norte. Organized in super blocks called superquadras, the basic style was a six- or seven-storey apartment building with elevators, along with three-storey walk-up buildings and a few ranks of row housing. The separation of uses might have been sketched out by Le Corbusier, although the housing was almost all lower than his "towers in the park."

Despite the vagueness of the plan—or maybe because of it—five years later the new city was open for business. Within ten years it had reached its maximum planned population of 250,000 surrounded by a wide greenbelt. Fifty years after that, the city was choked with cars and more than two million people lived beyond the greenbelt in more or less planned communities. The biggest employer in the area was the government, and people had to get to offices either in private cars, on buses, or on a Mêtro system that goes only a short distance into the countryside. Distances between government buildings were designed to be large so that the structures would stand out against the emptiness of the public concourses, with the result that the city is one where distances are usually too far to walk.

If you live in one of the superquadras, you probably wouldn't find the absence of concern for pedestrians along the Monumental Axis a problem. That's because each complex, like many of the sectors in Chandigarh, houses about 8,000 to 10,000 people and is served by a small shopping centre along with elementary schools and churches accessible on foot through tropical gardens. In theory, you wouldn't have to go outside your little neighbourhood for services. The problem—besides the fact that the neighbourhoods, planned to accommodate one car per housing unit, may now be faced with two or three vehicles per unit—is that less than a tenth of the Distrito Federal's population lives in the superquadras. The result is long, long queues of cars and buses transporting people who live on the periphery into the central city.

When I was there in the Brazilian spring of 2013, I tried walking the length of the grand concourse, but it was too hot and too far, so I took refuge inside one of the superquadras. There the air was cooler under the shade of mature trees planted when the complex was first built fifty years previously. A group of toddlers were playing in a small playground, while not far away an elderly woman sat watching. She was accompanied by a much younger woman, possibly a paid companion. The people who moved in here when the complex was built had a chance to buy the apartments in the 1990s, and many did. Now these original residents have aged, just as the buildings of the new capital have.

Therein lies one of the basic problems with concrete: even if structurally maintained, the material which was once thought to be a true "Rock of Ages," frequently doesn't weather well. A case in point: the outside of buildings such as Niemeyer's stunning cathedral needed to be painted or at least scrubbed down the day when I was there. Patches of greenish mould blemished it in shady places, like moss on the north side of trees in another climate, another hemisphere.

Time also has cast its shadow across Chandigarh and Brasília in another way: technology has moved on. As a young man, Le Corbusier wrote about cities as if his plans would be the very last word in urban development. He was all in favour of removing vestiges of the past and replacing them with new structures, new ways of living, with the implication that once corrected, once built properly, cities would be problem free in the future. At the same time, somewhat inconsistently, he had great faith in technology, and more than once called his Unité d'habitation a "machine for living." He could not—or certainly did not—imagine that around the corner lurked three inventions, three machines, that would seriously disturb the realizations of his ideas: air conditioning, computers, and wireless communication.

To walk today in either of the cities that reflect his ideas most clearly is to see just how much new developments transformed—or subverted—carefully considered urban landscapes.

Le Corbusier took care to include concrete sunshades—*brises-soleil*—on his buildings, even in temperate climates. The idea was to consider the angle that the sun shone down at different times of the year and to shield windows from direct sun during the hottest time of day. (Conversely in his plans for buildings at more northern latitudes, he also considered the best angles for letting in warming winter sunlight.) In Chandigarh, he and his team also designed pierced concrete block walls to allow ventilation and cooling breezes. But none of this was as effective in cooling buildings as is modern air conditioning. Air conditioning was not common even in North America or Europe when Chandigarh was built: in 1953 only a million air conditioning units were sold in the United States, and certainly few were available in a country where concrete itself was an expensive commodity.[107] The price of air conditioning has dropped drastically since then, however, while availability of electricity to run the units has increased greatly all over the world. Now wherever people can afford them, air conditioning units poke out from windows, or clutter up the balconies that Le Corbusier expected to be used by people seeking cool outside breezes. More recently, air conditioned malls on the edge of Chandigarh have been blamed for the decline of Sector 17, the former commercial district, because people are choosing to shop in newer, cooler buildings in peripheral districts.

The same holds true for many places in Brasília, although the public spaces I was in—museums, shopping malls off the Monumental Axis, hotels—all seemed to have more or less successfully integrated central air conditioning into their ventilation systems. Many office buildings, though, were peppered with window air conditioners, while satellite dishes and cell phone relay towers were nearly ubiquitous.

In Chandigarh, a series of photos taken between about 2012 and 2016 show the problems of integrating computers into underwired office buildings. Cables loop across offices and hang down outside, while stacks and stacks of paper file folders lie in seeming disorder on shelves pushed out of the way to make way for more modern equipment.[108]

If these are problems encountered in engineering smallish cities that were designed by some of the world's most renowned planners and architects, what will happen as China surges forward with its ambitious plans to rehouse so many of its millions and millions? What is just around the corner which could change the way people want to live? Can you engineer the future?

Who knows, but there is an important lesson in Chandigarh's development, according to Pradeep Singh, deputy dean at the Indian School of Business. "It wasn't just about taking an existing city and trying to make it grow, and make it better. This was a brave new leap of faith," he claims. "You took large tracts of farmlands and decided to make a whole new city, and you did it over a period of decades, according to a master plan. The needs of India are so large that that is the only way to meet our requirements for urbanization. Tinkering at the edges of existing cities, it is just not going to make it happen."[109]

The same may be true for China. The one thing certain is that China's plans for building housing, the road network, and other necessary infrastructure is resulting in and will continue to mean an immense demand for cement and concrete. Microsoft founder Bill Gates startled a lot of us a few years ago by making a much-repeated observation that between 2011 and 2013 China used more cement than the United States had in the twentieth century. To build modern America, the United States used 4.5 gigatons of cement between 1901 and 2000, while China, as it ramped up its housing and infrastructure offensive, consumed 6.6 gigatons in only four years.[110] In 2017, three years after Gates' reference point, Chinese domestic annual cement consumption reached a high of 2.62 gigatons, or about 70 percent of the United States' consumption over an entire century. Since then Chinese consumption

FIRE

has fallen somewhat, and in 2018 government agencies announced a reduction of about 10 percent of production capacity over the next few years. Business observers note, however, that this decrease will probably be matched by increases in cement production by plants owned by Chinese firms in Africa and the rest of Asia, where new cities are under construction in a half dozen countries.[111] The more cynical ones add that part of the projected decrease in cement production may be the result of a desire to export a pollution problem: cement and other industrial production was restricted in twenty-eight northern Chinese cities during the winter of 2017–2018 because of the terrible air pollution they caused.[112] But more about that in chapter 5, "Air."

What is clear is that Chinese cement production dips significantly every year for a reason that has little to do with the government trying to manage global expansion and everything to do with the very personal need of the Chinese—and everyone else—for a home.

Back in Roman times, Vesta was the goddess in charge of fire, that fundamental element.[113] The *Encyclopedia Britannica*, ever rational, explains that keeping a sacred fire burning was extremely important in a world where starting a fire was not an easy thing to do: remember there were no matches, not even a handy flint and steel combo to strike a spark back then. Therefore, the practical was combined with the spiritual in temples dedicated to Vesta, and in family shrines to her where a fire burned perpetually. The hearth where the family warmed itself and cooked its meals became over time the symbol of a basic unit of society: the family.

The Chinese pantheon is somewhat different. There is a god of fire, Zhu rong, but more important is the Kitchen God, who protects the hearth and family. The story goes that his wife records everything that is said in the family, and just before the Lunar New Year celebration in late January or early February, the Kitchen God goes to heaven, where he tells the Jade Emperor what good and bad things have been said and done in the family during the year. In preparation, delicious things are made to tempt him from reporting negatively, and the house is cleaned from top to bottom. Great quantities of food are cooked, and every year hundreds of millions of Chinese travel long distances to come home to feast and frolic. The link with concrete is that it is the material that builds the roads that carry the travellers home, it is concrete that made possible the jobs they toil at the rest of the year, it is concrete that is creating the millions of new kitchens in all those new buildings over which the Kitchen God will watch.

And because of this celebration, China's industries shut down for at least two weeks. So important is this holiday, so marked the disruption in the orderly way business is now done in China, that concrete production dips by up to 75 percent for the months of January and February.

The Roman Goddess Vesta might envy this kind of attention . . .

CHAPTER 4
WATER

ONE THIRD OF EVERYTHING IS WATER

The docks of the McInnis Cement plant jut out into the Baie de Chaleurs. As noted earlier, it was named by explorer Jacques Cartier because of the relative warmth of its waters. Located to the south of the Gaspé Peninsula, it extends west from Port Daniel-Gascons for more than 100 kilometres, and is divided into roughly two basins. The western one is relatively shallow, ranging from long inshore stretches no deeper than fifteen metres or so to depths of fifty metres maximum, which explains why that part of the bay is warmish. But the eastern basin, which begins where the bay widens out, is much deeper, and it's on the edge of it that the cement plant was built. The Gaspé current sweeps in cold, nutrient-rich water from the Atlantic here, making conditions good for marine life: Atlantic salmon, lobster, herring, crab, and cod have fed people along the coast for centuries and are still a major part of the region's economy.[1] More importantly for making cement and bringing it to market, the deep waters allow ocean-going ships to dock and deliver and pick up cargo.

There are beaches along this part of the coast. A little municipal park not far from the bed and breakfast where we stayed has trails down the red rock cliff that lead to a modest stretch of sand. It's possible that in high summer people might venture into the water to play, if not to swim, but we saw none during the late spring days that we were there, when the air as well as the water were very, very cold.

And certainly there was no way I was going in the water. I like my beaches much warmer, although my favourite place to swim in the whole world for a long time was one as intimately connected to cement as the shoreline at Port Daniel is now.

Millerton Lake is behind Friant Dam on the San Joaquin River in California. The swimming beach is about a twenty-minute drive from the neighbourhood where my in-laws lived in the northern part of Fresno. We would go there in the late afternoon on sizzling hot days, with the windows of the car wide open and the torrid wind rushing in. The route at that time went past new subdivisions before it broke out into agricultural land—grapes, peaches, a cattle feed lot. The sky seemed vast, even when it was discoloured with air pollution that collected in the long Central Valley of California.[2]

The valley floor itself is basically flat, tilting northward a bit, so that rivers and streams rising in the Sierra Nevada Mountains to the east ran toward a shallow arm of the ocean, San Francisco Bay. This estuary is where the San Joaquin River joins with the southward flowing Sacramento River. The two of them drain the interior of the state of California, carrying meltwater from the abundant snows that fall in the Sierra Nevada, the 1,600 kilometre long mountain range to the east.

No snow falls in the Central Valley in an ordinary year, and not very much rain: the average in Fresno is about thirty centimetres, most of it falling in the winter. Rain in summer, particularly late summer when farmers elsewhere may be praying for a break in a drought, is not welcome because it interferes with Thompson seedless grapes drying in the fields or on the vines to make raisins: 99 percent of all raisins produced in the United States come from Fresno County. But water is most definitely needed to grow the grapes, as well as the peaches, and the oranges, and the alfalfa, and everything else that has made Fresno County one of the most productive agricultural areas in the world.[3]

The connection between concrete and this abundance is one that most people find easy to make. I brought it up whenever friends asked me why I wanted to spend a couple of years working on a book about concrete. But they understood quickly when I began talking about how North Americans' diet would be fundamentally different from what it is today without the waters collected behind dams and then transported in concrete-lined canals.

Take, for example, a vegetable that my grandsons love and their grandfather detests: broccoli. Quite a bit is grown around Montreal, so it's no problem to find local broccoli in the summer and fall. But come November, what's available comes from California (about 90 percent of the production) or Arizona.[4] The seeds for the plants that show up here during the winter were sown the previous October in fields that are irrigated with water diverted from the big California dam projects. What we find in our markets in the middle of January and February is usually a 7.5 to 20 centimetre diameter bunch that has been hand-harvested and then fastened together with a rubber band before being placed in a waxed cardboard box weighing a minimum of 10.43 kilograms. The boxes are packed into refrigerated containers and then shipped the 5,000 kilometres to us, mostly along local roads and Interstate highways which have concrete as their base.

Although the broccoli grown here thrives on Quebec's regular late-summer rains (the plant doesn't like our hot midsummers), concrete is necessary for the California variety to get a chance to grow. The water that irrigates the broccoli from Fresno County that we eat comes from Friant Dam. It is one of several dams built to "move the rain." All were constructed during the 1930s and early 1940s in dry regions of the American Southwest and in Washington State on rivers that rise in faraway mountains. The dams served other purposes, among them providing work for the thousands of unemployed during the Great Depression. They also came to produce hydroelectricity that changed the course of history. Their recreational use—initially not considered important—nevertheless marked many, including me.

The construction of Friant Dam between 1939 and 1942 was the opening salvo in the struggle to remake the hot, dry San Joaquin Valley into an Eden, as then governor Earl Warren put it when the first water was sent into the canals of the Central Valley Project (CVP). The motivations that lay behind the CVP and the other big dams were complicated and many, including flood control, hydroelectric generation, irrigation, and to a lesser extent, providing water for people to drink. The repercussions have been enormous. Whether they have been good or ill is a question still up in the air.

Before we get to that, however, we should back up and consider how water, the ancient element, is essential to concrete itself.

Look at a video of a do-it-yourselfer mixing up a batch of concrete, and what you see to begin with is a pile or a pan full of powdery stuff (the cement) and a bunch of small rocks and sand. You can stir them together as much as you like, but you'll get nothing more than dirt. Just as you have to add milk to a cake mix to get a batter that might be worth something, you have to add water to the dry ingredients to make concrete. That's when the magic begins. The water sets off the chemical reaction that will turn the slurry into rock.[5]

Almost any kind of water that you can drink will do. So will untreated water from lakes, rivers, or streams. Even clean seawater is all right for concrete that doesn't contain metal reinforcing, but water with a heavy silt load or containing contaminants like sulphate and alkalis over certain limits can adversely affect the strength of concrete. So can adding too much water to the mix: usually concrete shouldn't be more than about 30 percent water at the beginning of the process. As explained in chapter 2, on earth, what happens when concrete sets is that the chemical reaction produces a latticework of crystal-like material that binds the aggregate and sand together and which strengthens over time.

Concrete has to be poured within ninety minutes of water being added. If not, it sets with pretty dramatic consequences. Many a do-it-yourselfer has learned this the hard way when trying to clean up after taking too long to smooth a concrete slab. Construction professionals know this all too well too, which is why having a ready-mix batch plant near a building site is important and why delays in traffic can spell disaster. Back at the Concrete Basics seminar I attended at the World of Concrete trade show, the slide that got one of the biggest reactions showed blobs of hardened concrete that had been forcefully removed from a concrete mixer truck when it couldn't be emptied on time. To me it looked like hardened lava after a volcano eruption. To the concrete guys present, it must have looked like thousands of dollars in time, equipment, and material wasted. To be sure, adding more water can slow things down, but it may also produce weaker concrete, the seminar leader warned.

If the mix of cement, aggregate, and water is right, however, after twenty-four hours (in most cases) concrete is strong enough for the forms into which it is poured to be removed. About seven days is required for the concrete to get to useful strength, however, so concrete work is a balancing act where timing is very important. The strengthening process continues even after the poured concrete can bear weight: the process goes on sometimes for years. The heat produced in the process also takes a long time to dissipate. For big projects like the Hoover and Grand Coulee dams—which we'll consider more closely later—the cooling time could be tens of years unless it's speeded up by external forces like passing cold water in pipes through the poured section.

There are a number of ways to test how strong concrete is, but nearly all can only be done after the fact. These include taking a core from fresh concrete, letting it cure, and then compressing it to see how much weight it can bear. If the results are bad, replacing the defective sections can be very expensive. But that test requires waiting days, and while it's frequently done, builders rely on one quick test to determine if the batch being delivered is likely to pass muster.

What you do is pour a sample of the concrete into a flowerpot-like receptacle, turn it upside down, and then remove the pot. The material will slump some. How much depends on the water and cement content, among other things. Too much slump (and 30 percent is a standard measure) indicates that the concrete will cure with problems. It will not be able to bear weight adequately or it will be more susceptible to deterioration. This is because tiny channels are formed in the concrete as excess water migrates to the surface. The concrete becomes more porous to water infiltration, letting in damaging chemicals.

On nearly every construction site, the guy (or gal) accepting a truckload of ready mix will do a slump test before signing off on it. Sometimes there can be a lot of pressure to let a questionable batch through. A friend who had a summer job on a big construction site decades ago still vividly remembers turning down three loads in a row which didn't meet the slump standard, and then being summarily removed from the job because "we don't operate that way." Not surprisingly the project afterwards was fraught with structural problems, and parts of it were subsequently demolished.[6]

So, while water is essential for making concrete, it can become the poisoned elixir that destroys it, particularly if the project is poorly designed. Fissures are nearly universal in concrete, and part of the skill of working concrete is controlling where the fissures occur. Those joint lines you see in sidewalks are put there in an attempt to guide the cracks. But for various reasons having to do with the cement formulation used, pores and fissures in the wrong place can allow water to percolate into the concrete, sometimes causing a lot of damage.

A case in point is one close to where I live now: the Champlain Bridge across the St. Lawrence River that we spied from Habitat 67. Completed in 1962 when "we were in a great hurry to build things," the bridge was terribly designed, asserts Saeed Mirza of McGill University in Montreal, who spoke out about the bridge's problems for years before authorities agreed that it should be replaced. Prestressed concrete—where, as mentioned before, steel elements are stretched and then embedded in concrete which is poured around it—was the basic material used. The technology hadn't been used in Canada before, and a more traditional technique of using steel girders and concrete would have been more lasting, Mirza says. If that had been the case, a corroded section could have been rather easily and quickly replaced.

Furthermore, the steel inside the bridge's concrete wasn't coated to protect it from attack by salty water created by de-icing salt, and when repairs were attempted, they were badly thought out. Some steel reinforcement elements were installed on the exterior of the structure, and were protected by plastic tubing. That might have helped against salty water, but because the tubing breaks down in the presence of sunlight, the fix was far from adequate. There also was a basic design flaw: Mirza says that the drainage was awful, without even downspouts. Salty snow meltwater passed over a parapet wall and then ran down the bridge's piers. The result was disastrous for the bridge, although,

Figure 4.1: The Champlain Bridge: damage done when bad design meets de-icing salt and runoff.
Photo: Mary Soderstrom.

to be fair, the widespread use of de-icing salt was not anticipated when the bridge was designed.

The expectation then was that snow would be removed with ploughs and that cinders would be spread on the surface to provide traction. Back in those days when coal was still used for heating in many buildings, the abrasive residue was easy to obtain. But three things effectively removed the use of cinders from consideration. The first was the move away from coal for heating toward oil and gas, for environmental reasons. The second was the discovery that cinder was a valuable additive to concrete itself, either for its pozzolan qualities or as a lightweight aggregate. But the third and probably the most important came as fallout from a study done in Germany, which showed that using salt to melt snow on roadways resulted in far fewer traffic fatalities.[7] Jurisdictions all over North America turned to salt for road de-icing almost immediately, and by 2014 the United States was using sixty-two kilograms of de-icing salt per person, even though in a good half of the country, the climate made road ice rarely a problem.[8] Mirza says that by the 1980s four to five times more salt was being used on the Champlain Bridge than was needed.

By 2009 it was clear that it was going to cost far more to rehabilitate the bridge than to build a new one—and that the need to do something was urgent because the bridge, which handles 156,000 vehicles a day—fifty million vehicles a year—was literally falling to pieces.[9] An engineering report made for federal authorities in 2011 did not rule out the possibility of "partial or complete failure"—that is, collapse—even after c$158 million was spent to temporarily reinforce the bridge.[10]

Mirza and colleagues compared the bridge with one standing only a couple of kilometres away: the Victoria Bridge, built in 1860.[11] Originally a railroad bridge, but repeatedly upgraded and maintained, this steel structure is good for another sixty to seventy years of service. The old Champlain Bridge, on the other hand, is being dismantled following the opening of the new Samuel de Champlain Bridge to traffic in 2019. (Yes, the name is almost the same, and its choice prompted much discussion, but that's not a story for now.)

The need for demolition points up another problem with poor concrete construction: what to do with the millions and millions of tons of concrete rubble produced each year? In the United States, concrete made up 70 percent of the 534 million tons of construction and demolition waste produced in 2014.[12] In the case of the Champlain Bridge, the problem is compounded because, not only must the old structure be torn down, but building the approaches to the new bridge meant the demolition of several freeway overpasses that had been rebuilt only a few years before. Some of the construction waste can be recycled, but far from all. Reconstruction of another nearby freeway—also necessary because of badly aging concrete—is reusing concrete and asphalt in embankments for new access roads, while the steel rebar is being recycled by a private company.[13] The consortium that built the new bridge said that steel and concrete from the old bridge will also be recycled and that local recycling firms should have the capacity to handle the debris. Nevertheless, the process will cost at least c$400 million and take three to four years. That makes for quite a load of garbage.[14]

But one shouldn't read into the Champlain story an argument against using concrete for major public works projects like bridges, Professor Mirza says. No, what's necessary, he says, is to get away from the engineering credo that is far too prevalent of "design it, build it, forget it."

In other words, the implications of projects built with concrete should be pondered carefully, both for the long- and the short-term. When that is not done, the repercussions can be enormous. With a bridge it can mean loss of life if the structure fails, or at the very least enormous public expenses: when Genoa's Morandi Bridge—built about the same time as the first Champlain Bridge and with some of the same structural problems—collapsed in 2018 it killed forty-three and left 600 homeless.[15]

That's bad enough, but problems in the construction of big dams could carry even worse consequences were a dam to fail.

MAKING THE DESERT BLOOM, KEEPING THE FLOOD AT BAY

As Joan Didion wrote in her book of essays *The White Album*: "Some of us who live in arid parts of the world think about water with a reverence others might find excessive."[16] I am one with Didion. San

Diego, where I spent a good part of my childhood, gets slightly more rainfall than Friant does in a year, about thirty centimetres, but that is definitely not enough to support a population that reached half a million by 1960 and two million by the end of the twentieth century. When I was growing up, the water we drank was piped in from the Colorado River, and its taste was something that made newcomers turn up their noses or even gag. Even then, we had low-water plants as landscaping, and I was taught to be fanatic about turning off the water when brushing teeth. No grass for our school playgrounds, of course: my memories are of constant skinned knees and elbows from collisions with the asphalt paving.

There was water nearby, but it was the salty ocean. We spent a lot of time at the beach, but swimming there was nothing like swimming at Friant. First of all, the water was much colder. Most of the time the California coast is affected by the California Current, a swooping river of ocean water that chillily runs southward from the Gulf of Alaska and British Columbia along the eastern edge of the Pacific Ocean basin.[17] Usually it only begins to warm up when it reaches the Baja California peninsula, which means that water temperatures at Southern California beaches are as much as ten degrees cooler than those on beaches at the same latitude in the eastern United States.

Then there are the waves. My high school was one where guys (and they were almost all guys back so long ago) hung out the windows of the second floor every morning to see what the surf looked like. If it was good, they'd be gone after lunch. I caught a few waves myself, bodysurfing. Swimming, though, was hard unless you fought yourself out past the line of breakers.

The calm waters of Millerton Lake behind Friant were quite different. There I could paddle around wearing a sun hat and prescription sunglasses, something you'd never do at the ocean. But with my specs on I could actually see across the lake to the golden hills on the other side, even glimpse the still-snow-capped Sierra Nevada mountains off to the east. The only real problem was the way the shoreline fluctuated as the season progressed and more and more water was drawn down to irrigate the crops that would feed millions. By August, what in the spring had been an island in the lake had become a hill, and you had to drive several kilometres farther along the shoreline to get to the area reserved for swimming.

The water of the 245-kilometre-long Friant-Kern canal—visible at a couple of places on the road to the recreation area—never diminished that I could see: the irrigation water in the canal was the real raison d'être for the dam. Without it the billions of dollars of fruits and vegetables grown along the canal would wither and die. Fresno County alone grows more agricultural crops than twenty of the fifty states in the United States. According to the most recent statistics, in 2018 it produced nearly US$1.2 billion worth of almonds, US$1.1 billion worth of grapes, US$862 million of pistachios, and more than half a billion U.S. dollars of poultry, not to mention US$15 million worth of broccoli.[18]

It was that kind of largesse that those who dreamed up North America's irrigation projects had in mind. Wallace Stegner, in his novel *Angle of Repose*, vividly portrays it when he describes the promise of an early irrigation project on the Colorado plateau:

> Eventually willows and lombards will line the [canal] from the canyon mouth to its lower end and . . . drop their leaves on its current to spin in its slow whirlpools and snag on weeds and roots and make a resting place for darning needles and dragonflies. By their living green presence along the line of the ditch they will be . . . the truest testimony to the desert's fertility, and the beacon of hope to settlers and their families. This is all in that future when our grove will be tall and cool around our house, and when we will leave that coolness for a different coolness on the banks of the Big Ditch, . . . and watch the sunsets reflected in our man-made river sixty feet wide.[19]

Stegner's hero wasn't successful in irrigating the high, dry land where the Colorado River begins its run southward, but others were.

Scarcity of water wasn't the only thing that prompted the building of these dams, however. Although it seems almost contradictory in regions of little rainfall, flood control initially was perhaps an even bigger incentive.[20] Torrential rains or rapid snowmelt in the Sierras meant that far more water surged out of the mountains than could be contained in riverbeds. For Didion, born in 1934 and so a little girl when many of the big dams were under construction, the California winters of her childhood were marked by "all night watches on rivers about to crest, by sandbagging, by dynamite on the levees and flooding on the first floor."[21]

Even the Colorado River, which most years now ends unceremoniously in stagnant, saline puddles, has run rampant several times since the country it passes through was settled by Europeans. It rises in the Rockies, and over its nearly 2,400 kilometre course runs through some very dry territory. Floods on it can be due to high snow and winter rains in the mountains which produce crests in spring time, or because of summer thunderstorms during what locals call the "monsoon" season. One of the most spectacular floods occurred in 1905 when sixty-five kilometres of the main line of the Santa Fe Railroad was destroyed in the rich farmland of California's Imperial Valley. The river changed course, flowing into a sink that is below sea level, and forming what is now called the Salton Sea.[22] Pressure mounted in following years to even out the boom and bust river cycle: the Hoover Dam was promoted by farmers downstream as a way to avoid being flooded out in some years, and dried out in others.[23]

Building a series of dams on the Colorado hasn't solved all the problems, however. They had no effect on the Big Thompson Flood of July 31, 1976, which occurred on the upper reaches of a Colorado tributary. Called one of the most devastating in United States history, it happened when between 30 and 35 centimetres of rain fell in 4 hours, sending a 6-metre-high wall of water down a steep, V-shaped canyon. Some bodies of hapless vacationers enjoying a summer holiday weekend were carried 40 kilometres downstream. It is only because of the rather remote area where the flood took place that no more than 143 deaths were recorded.[24]

In fact there are only three ways to counter floods. The first doesn't depend on concrete: it is to live with them. That is what the Egyptians did for the most part: the annual floods on the Nile brought loads of silt and essential water to their crops. Regular floods can be beneficent: for several thousand years the flooding rivers that rose in California's Sierra Nevada laid down their burdens of silt, creating the rich land that now makes the Central Valley so fertile. A modern corollary is to allow flooding rivers to spread out in wetlands and marshes in specific places so that the water can be absorbed by the soil and so replenish underground water.

That no-concrete strategy is considered a non-starter in most places today because in built-up areas, houses constructed on flood plains get— surprise, surprise—flooded regularly. You can plan for this, though: the Brazilian city of Curítiba on the Paraná River has forbidden residential or commercial development in certain flood-prone areas. Instead they are part of the parks and recreation system, where the only construction allowed is more or less flood-proof concrete recreational structures like picnic shelters.[25] (Of course, this only works up to a point: in 2019 Curítiba experienced some of the worst flooding in decades, largely, say environmental experts, because run-off is much greater than it was previously since much more of the land is now covered in concrete. But more about that a little further on.)[26]

The second way to deal with floods is to try to channel them, by building levees and dikes designed to keep water flowing quickly to the nearest ocean. That's the strategy used along the lower Mississippi, with mixed success.[27] The river drains 41 percent of the continental United States and parts of Manitoba and Saskatchewan. When flooding occurs on the rivers which feed into it, the volume of water can be enormous. Levees have been built on the lower Mississippi since the French tried to protect New Orleans from floods in 1718. At the beginning they were basically wide berms of earth that paralleled the river's channel. In recent years, as the levees have had to be built higher and higher, concrete flood walls have played a big role in containing the river where it runs through cities.[28] Yet in years of big floods, the levees are frequently breached. Some are planned breaches to allow flood waters to invade low-lying land and avoid worse damage in more built-up places, but the prospect looms of more "100 year" and "500 year" floods following greater rainfall as the climate changes.[29] In addition, as seen in Curítiba in 2019, when it rains hard, concrete multiplies the impact of downpours because paved roads and parking lots prevent rainwater from being absorbed by the soil. Instead, excess water is channelled into storm sewers and back to the rivers.

To hold back floodwaters, levees on the Mississippi are now being built as high or higher than a three-storey building. The one in Cape Girardeau, Missouri, north of St. Louis, was built to withstand a flood of 16.75 metres. Rising water hasn't reached that benchmark yet, but the levee and flood wall successfully held during a fifteen-metre flood in 1993. In St. Louis itself, which is just south of where the Missouri joins the Mississippi, I remember driving along the levee, thinking that I'd get a good look at the mighty river. But I was disappointed because in most places the roadway was several metres higher than the river was that day, with few spots where you could get down near the water. Even the park surrounding St. Louis's signature monument, the Gateway Arch, drops abruptly toward the river. Most of the time the great flight of stairs leading down to the concrete flood wall is a favourite challenge for physical fitness buffs, but in June 2019 they had to change their routines as water lapped at the topmost steps.

However, impressive as it is, the levee system on the Mississippi doesn't compare in age or grandeur with that of the Yellow River. The river, China's second longest, is sometimes considered the cradle of the ancient civilization which has flowered along it for millennia. But it's also called China's Sorrow because the river's wilfulness has killed literally millions.

The river rises in western China, in mountains north of the Gobi Desert. At its beginnings it runs clear, but when it descends to cross the relatively dry North China Plain it picks up a burden of loess, those extremely fine particles of earth that are light enough to be blown by the winds for long distances. The yellowish-brown sediment gives the river its name as it crosses the wide, gradually sloping plain that is much of eastern China. This load of silt is responsible for the plain's fertile soil, but when combined with the slow descent of the river, the stage is set for catastrophic changes in course that make the San Diego River's switch from one bay to another, or the Colorado's adventure to the Salton Sea look like mere twitches.

To keep the river in its bed, over the centuries the Chinese built levees that in places are metres higher than the surrounding land. The result is that in non-flood time the river actually flows above the land it waters. When the silt in the water builds up at a natural obstacle, the water seeks the path of least resistance around it, sometimes overtopping the levee, other times simply eroding it away. When a breach occurs, the river's waters pour out, unable to return to their previous channel. They adventure widely, suddenly and disastrously flooding towns and agricultural land.

Over the last 2,000 years, it's estimated that the river flooded about 1,500 times.[30] For several centuries it ended in the Yellow Sea to the south of the Shandong Peninsula, the one which juts eastward from the middle of China's east coast toward the Korean Peninsula. But in 1852 the river shifted almost 300 kilometres to the north so that it flowed into the Bohai Sea, an embayment on the north side of that peninsula. How many people were killed in the floods and famine that followed is unknown, but estimates place the number at hundreds of thousands. Then an even worse flood event occurred in 1931, when the river shifted back to the south: estimates of the deaths at the height of the flood and in the months that followed due to disease and hunger range from 85,000 to four million. Seven years later, Chinese Nationalist forces attempted to turn back Japanese troops by destroying the dikes. When they were rebuilt, the river was diverted to the north. It now once again flows to the north of Shandong Peninsula, entering the Bohai Sea about 200 kilometres south of Qinhuangdao, where Moshe Safdie is building the Golden Dream Bay development.

Which brings us to the third way to control rivers: by building dams.

Of course, people have been building dams for a long time. Small dams—some just big enough to store the waters of a small stream, or to channel floodwaters—were an integral part of early, very successful irrigation systems in Mesopotamia and China. The impact of these projects should not be underestimated, but the first big dam we know about was constructed in Egypt.[31] It was built about 2600 BCE about thirty kilometres south of Cairo and was designed to protect a valley settlement from Nile floods. Built of two kinds of limestone with a core of rubble, it appears that, before the dam was completed, a flood swept over the structure and that part of the core was eroded away. As the authors of an article on the dam note, because it was such an early structure, dam builders did not have much experience to draw on in building this "gravity" dam, where the sheer weight of the structure is supposed to keep back the water.[32]

Dam builders learned how to solve these problems, though, and the vestiges of several other gravity dams dating from near the turn of the Common Era remain. Not surprisingly, given the Romans' mastery of concrete, two dams built by them are still functioning today, both near Mérida in Spain. One, the Proserpina Dam, is built of earth and faced with Roman concrete, while the other is a gravity dam made with stones and clay and also faced with Roman concrete.

Gravity dams from the beginning were developed for locations where a river was to be impounded in a relatively wide valley. But by the thirteenth century CE, engineers discovered that they could use the same principles that made arches in buildings possible to create arch dams. In them, the weight of the dam itself has a role to play, but it is aided by the water, which also pushes the dam structure against the sides of a narrow gorge the way an architectural arch distributes its weight to the sides of a building. Not only do arch dams usually require less material to build, they can be extraordinarily beautiful.[33] Take the dam built in the thirteenth century at Kebar in central Iran. Still standing, it is twenty-six metres high and fifty-five metres long at the crest, and is amazingly well preserved. Now nearly 800 years old, it has no cracks or slippage, a tribute to the Mongol engineers who planned it, and to their use of a lime mortar to which they added the ash of a desert plant that seems to have acted as a natural pozzolan.

Run-of-river dams may combine features of gravity dams and arch dams but they differ from dams built primarily for flood control or irrigation in that they do not have very large reservoirs and rely on a steady flow of river water year round. Chiefly used for generating electricity, examples are the Chief Joseph complex (which also has a large irrigation component) on the Columbia River downstream from Grand Coulee, and the Beauharnois one on the St. Lawrence near Montreal.[34] Chief Joseph impounds some water, but the run-of-river dam on the St. Lawrence in no way tries to tame the river which drains nearly as much of North America as the Mississippi does.

Over the last seven decades since China's great civil wars ended, resources have been freed to build infrastructure that, among other things, has more or less brought the Yellow River under control. Today, on the upper and middle reaches of the river there are 173 large- and medium-sized water retention reservoirs, with two large flood retention basins farther down.[35] In addition, sixteen hydropower dams were built between 1960 and 2016. The river now seems conquered, although reduced rainfall throughout its watershed and greater withdrawing of water probably have also contributed to its reduction to manageable proportions.

When working on this book, I had no chance to visit a Chinese dam, but I did see two of the iconic American ones: the Hoover and Grand Coulee dams. One of the activities offered during the World of Concrete trade show was a guided tour of Hoover Dam and the graceful Mike O'Callaghan-Pat Tillman Memorial Bridge that flies from one side of the Colorado River canyon to the other just downstream from the dam. Both are wonders of concrete construction, and the folks who filled the two buses on the field trip listened carefully to what the guides had to say and took in what we were shown with the eyes of professionals. Among the things we learned was that Las Vegas was nothing more than a sleepy railroad junction town before the Hoover Dam project was launched, and that railroad spurs to bring in construction material were built from what even then was sometimes called Sin City. A company town called Boulder City was set up about forty-eight kilometres from the dam site, where alcohol, gaming, and prostitutes—big stuff in Nevada already back then—were not permitted. The idea was to keep the workers who lived there on the straight and narrow, and ready to report for work when their next shift rolled around.

The cement for the dam came from the California Portland Cement Company plant at Colton near Los Angeles. Electricity to power operations on-site also came from Los Angeles. The flow of current was reversed after the dam was completed, so that Colorado River hydropower lights Los Angles today. Nearby Las Vegas is off that grid, though, and gets most of its juice from gas turbines and solar panels.

As for the O'Callaghan-Tillman bridge, it was built with concrete from a nearby, specially built ready-mix plant where both silica fume and fly ash were added to the mix for workability and set times. Placing the concrete was frequently a race against time, wind, and bad weather: with temperatures in the mid- to high thirty degrees C, care had to be taken that the mix didn't start to set in the trucks.

The highway that the new bridge carries, U.S. Route 93, is the main road from Flagstaff, Arizona, to Las Vegas. It used to run across the top of the dam as a two-lane roadway, but security concerns in the early twenty-first century—commercial truck traffic was prohibited after the 9/11 terrorist attack, in fact—led to the construction of the new bridge that was opened in 2010.

But the bridge is not just an exercise in danger avoidance. Care was taken to make the bridge's arching structure—it's the widest concrete arch in the western hemisphere—echo the arch of Hoover Dam. In the late afternoon when shadows are long, the harmony of the two structures is evident.

That's fitting because the dam itself is full of art deco details that reflect the place it was designed to take in the American dream. The entrance gallery—built mostly of concrete, of course—has the curving polished brass of a 1930s movie palace. The powerhouse where the turbines are has mosaics on the floor which represent water and what used to be called Native American themes. Along the road that crosses the dam itself, the buttresses are decorated with bas reliefs in the muscular style frequently used in public architecture during the Franklin Delano Roosevelt (FDR) administration and the troubled years of the Great Depression. Many projects built then aimed both to provide work and to build a better world.

FDR was elected president in 1932. He was swept into office by a wave of dissatisfaction with Herbert Hoover's administration, which did not effectively counter the problems created by the worldwide economic crisis that followed the stock market crash of 1929. What would become Hoover Dam was already in the works, however, and Hoover, who was a trained engineer, had championed the project. Roosevelt,

Figure 4.2: The Hoover Dam on the Colorado River is perhaps the best known of the many big dams constructed in the United States during the 1930s and 1940s. Photo: Mary Soderstrom.

once in office, ordered full speed ahead on it, but took care to change the name to Boulder Dam after the canyon in which it was built. It was only after another Republican president was elected twenty years later that the dam's name was changed back to Hoover.

By the mid-1930s, this dam and others under construction offered the promise of work in a time of very high unemployment, but it was work that took a toll. The ninety-four men killed during the dam's construction are memorialized in a striking bas relief of an idealized worker with the legend: "They died to make the desert bloom." But these "official" deaths were not the only ones. It was a time when safety measures were sketchy at best: hard hats were required, but the desire to get the work done as fast as possible led to the use of gasoline-powered trucks and equipment deep inside the dam. Many casualties didn't occur on the job, but later after the men who'd been overcome by carbon monoxide had been taken out. Legend has it that some men who knew their time had come crawled to the Arizona side of the dam because death benefits were better for people who died there than on the Nevada side.[36]

The truth is that the big North American dam projects employed a relatively small number of the unemployed. To be sure, 12,000 men and women worked at Grand Coulee on the Columbia at one time or another between 1933 and 1943, while Hoover Dam employed more than twice that with a monthly average of 3,500. But remember that in the 1930s, 2.5 million people were pushed off their Dust Bowl farms by drought or thrown out of factory work as the bottom fell out of the economy. The very possibility of work brought men—many with their families—surging into the project zones.[37]

John Steinbeck's novel *The Grapes of Wrath*, published in 1939 but finished in 1933, gives some idea of how desperate the job seekers were. He tells the story of the Joads, who bear some resemblance to the family whose misadventures with cement were chronicled by William Faulkner in *As I Lay Dying* (see chapter 2, "Earth"). The Joads, however, have had enough of rural poverty and so set off for California from Arkansas with the idea that they'd find a piece of land to farm, and with it, peace and prosperity. They travel Route 66, that "long concrete path across the country" which is:

> the path of a people in flight, refugees from dust and shrinking land, from the thunder of tractors and shrinking ownership, from the desert's slow northward invasion, from the twisting winds that howl up out of Texas, from the floods that bring no richness to the land and steal what little richness is there. From all of these the people are in flight, and they come into 66 from the tributary side roads, from the wagon tracks and the rutted country roads. 66 is the mother road, the road of flight.[38]

When they get there the men are offered work digging a trench for a concrete pipe, but they soon discover that the wage has dropped from thirty cents an hour to twenty-five cents an hour. Nevertheless, they take it:

> Timothy looked at the ground. "I'll work," he said.
> "Me too," said Wilkie.
> Tom said, " . . . Sure, I'll work. I got to work."
> Thomas pulled a bandanna out of his hip pocket and wiped his mouth and chin. "I don't know how long it can go on. I don't know how you men can feed a family on what you get now."
> "We can while we work," Wilkie said. "It's when we don't git work." [39]

A twenty-first-century corollary might be seen in China today, where people are on the move from poor rural areas to places where conditions look better. But unlike China's millions of units of new concrete housing, there were no new ready-built cities waiting for the Joads.[40] Nor were the Joads, unlike the lucky ones who were at the right place at the right time when the dam projects were hiring, able to find good paying jobs: Tom and Timothy and Wilkie only got two bits an hour, but the unskilled labourers at Grand Coulee were paid at the princely rate of seventy-five cents an hour.[41]

Nevertheless, the Joads, and people like them, kept going on hope. They were probably not untouched by the hype surrounding the dams, which were called the biggest things since the pyramids, a symbol of what men can accomplish if they work together, projects that would make the land bloom.

The spin was helped by folk singer Woody Guthrie, who was hired by the U.S. government, for US$266.66, to spend the month of May 1941 writing songs about the Grand Coulee and Bonneville dams. The idea was that his music would counter attempts by private power interests to stir up opposition to the projects.[42] It did, and even today when one has to recognize that the dams have their downside, Guthrie's songs "Roll on Columbia" and "Grand Coulee Dam" are captivating.

What is more, at least one of the twenty-six songs he turned out that month was prescient. "Pastures of Plenty"—recorded subsequently by many other singers, including Bob Dylan, Harry Belafonte, and Alison Krauss—tells what would come after the hardscrabble migrants worked the now irrigated land: they would defend with their lives the fields that they worked so they can "always be free." Seven months after the song was written the United States would enter World War II, and thousands of young men would join the military, while thousands and thousands more would find work in defence industries. The war would end the Depression, but that's a topic for another book.

So, the dams provided flood control, electricity, and water for agriculture and domestic use: that was their reason for being. But one of the consequences which is underappreciated is the role they played in setting up one of the better aspects of health care in the United States. Kaiser, one of the companies in the consortium that built Grand Coulee, established the first successful prepaid health-care insurance plan to care for workers on the project. The plan became Kaiser Permanente, the largest managed health-care organization in the United States, which insured 12.3 million Americans in 2019. Its model of up-front contributions from the employer was the brainchild of Sidney Garfield, a young surgeon who had worked on the Los Angeles aqueduct project, and who had a revolutionary take on providing health care. He's quoted as saying: "To the private physician, the sick person is an asset. To us, the sick person is a liability. We'd go bankrupt if we didn't keep most of our members well most of the time."[43] In a country where getting universal health care has been a struggle for decades, the legacy of Kaiser's insurance for workers is a welcome bright spot.

What the dam projects didn't do, though, was support the American (and Canadian) ideal of the small family farm. The Joads had seen the beginning of that, as Steinbeck writes:

And it came about that owners no longer worked on their farms. They farmed on paper; and they forgot the land, the smell, the feel of it, and remembered only that they owned it, remembered only what they gained and lost by it. And some of the farms grew so large that one man could not even conceive of them any more, so large that it took batteries of bookkeepers to keep track of interest and gain and loss; chemists to test the soil, to replenish; straw bosses to see that the stooping men were moving along the rows as swiftly as the material of their bodies could stand . . . . And the owners not only did not work the farms any more, many of them had never seen the farms they owned.[44]

When the Columbia River project was first being talked about, the public rhetoric had it that yeomen farmers and their families would be the beneficiaries of big dam irrigation. It was estimated that when the river's waters were tamed and channelled, the 80,000 acres of the Columbia basin could support 80,000 people and 10,000 farms. Instead, in 1973, twenty-five years after water began to be pumped from the dam over the hill to where it could be sent by gravity to waiting crops, there were only 2,290 farms. The following year a study found that 5 percent of the farmers received 20 percent of the benefits. The Joads probably wouldn't have found that surprising, given the way their experience showed so many things that were stacked against the little guy.

This wasn't a terrible subversion of the dam-building project, however, according to historian Paul Pitzer, who wrote an exhaustive book on Grand Coulee. Had the original planners achieved what they said they wanted, Pitzer wrote in 1994, there'd be "a collection of family farms ranging from forty to eighty acres, none of them capable of supplying their owners with a satisfactory living. The area would be a rural slum."[45] Instead what we have are orchards and vineyards, abundant

Figure 4.3: Irrigated orchards on the Columbia River, not far from the Chief Joseph Dam in Washington State. Photo: Mary Soderstrom.

fruit and wines that win international prizes in Europe, all of which would have been unimaginable before the dams on the Columbia.[46]

Grand Coulee Dam today is a bit off the beaten track. To get there you have to take Washington 155 or 174, connecting from U.S. 2 or U.S. 395. Interstates 97 and 90 are even farther away and it takes about an hour and three-quarters to get to the nearest big city—Spokane, population 220,000—by car. So you approach the dam by driving for kilometres over prairie and dry hills interspersed with luxuriant groves of irrigated fruit trees that provide apples, cherries, and pears to a continent. Rural poverty is not apparent, except possibly in places like the small town of Othello, where a neighbourhood of trailer homes appears inhabited by seasonal agricultural workers.

A good part of this territory is the ancestral tribal lands of Indigenous Peoples who belong to what are now called the twelve Colville bands: the Methow, Okanogan, Arrow Lakes, Sanpoil, Colville, Nespelem, Chelan, Entiat, Moses-Columbia, Wenatchi, Nez Perce, and Palus. Like other Indigenous Peoples, they suffered in the nineteenth century when European settlers arrived. But the coup de grâce that killed their way of life came when Grand Coulee blocked all migration of salmon to the upper Columbia.

Salmon are anadromous fish, which means that they pass part of their lives in fresh water and part in the sea. Young salmon leave the swiftly flowing streams where they hatched to descend rivers to the ocean. As adults they return to their home river and fight the current upstream to their home stream before spawning and then dying. In the past, as many as seventeen million salmon made the run up the Columbia each fall, reaching as far as the Columbia Lakes in Canada, 2,000 kilometres from the mouth of the river. Indigenous Peoples made the fish a major part of their diet for as long as 10,000 years.[47]

Then came the dams. The ones lowest on the Columbia were built with fish ladders which were supposed to resemble the rapids through which the fish once successfully navigated. To some extent that has worked, although runs have been reduced to about two million. But the fish spawned on the upper Columbia were stopped at Grand Coulee because no fish ladders were installed when it was built. Nor were they added at Chief Joseph Dam, a more recently built one somewhat downstream on the Columbia's meandering path to the sea. This has meant big changes for the Colville bands who now live mostly on the Colville Reservation, north and east of the Grand Coulee Dam site. They govern themselves, raise horses, and work for enterprises and governments around the area.

We didn't see a hint of that when we arrived because we came toward the dam from the south at the end of a hot afternoon. The river flowed from the east, however, which meant that, frazzled and impatient, we saw Lake Roosevelt, the water backed up behind the dam first, and not the structure itself. The effect was considerably less exciting than the approach to Hoover Dam, where the combination of the arching bridge and the arching dam set in its deep gorge is as harmonious as

Figure 4.4: Even bigger than Hoover Dam, the Grand Coulee Dam on the Columbia River provided electricity for war industries during World War II, and for making the plutonium necessary for two of the first atomic bombs. Photo: Mary Soderstrom.

the inside of a Gothic cathedral. Grand Coulee, on the other hand, is massively horizontal and much bigger. It took nearly four times more concrete to build than Hoover—12 million cubic yards compared to 3.25 million cubic yards—and at the moment it generates far more electricity, 6.48 gigawatts a year to Hoover's 4 gigawatts.[48]

Early the next morning, though, we found another vantage point where the view did Grand Coulee justice. The sun had not risen too high above the bluffs on the other side of the river, and the air was still cool. The whole dam was visible from the hilltop, with the blue water of the reservoir echoing the paler blue of the sky. The big scar where aggregate had been mined to make the concrete for the dam still was there, but we didn't spy any remains of the railroad line that had brought cement from the big plant in the town of Concrete on the other side of the Cascade Range as well as other material to build the dam. Dry grass and dusky green brush covering the hills contrasted with the billows of reddish-brown lava that had poured out in some convulsion of the earth ages ago. The wind blew and two big birds—eagles, perhaps—soared on the awakening thermal updrafts.

We looked and looked. The sign said this was the Crown Point Overlook State Park, but it didn't say who was responsible for the round concrete structure where we stood. The pillars that held up the circular roof could have been inspired by a Roman temple. The central opening through which one might see the stars at night made me think of the opening in a ritual building closer to home, a round house or lodge built by Indigenous people, or, going farther afield, the opening in the centre of the grand dome of Rome's Pantheon, the *occulus*. Whatever the architect's inspiration, the place seemed that morning to transcend the ordinary, to make a connection with things as elemental as earth and fire, water and air.

ENERGY FROM WATER

But the dam was connected to the wider, more mundane world too. Down below we also could see the rows of high-tension pylons carrying electricity from the dam to the homes and farms and industry of the Pacific Northwest. Grand Coulee is the biggest hydroelectricity producer in the United States and one of the biggest in the world.

For a long time I thought that hydroelectricity sounded like a contradiction in terms. It's not immediately obvious how you can get one from the other, especially when all the safety signs warn to keep electrical appliances away from water, for fear of electrocution. But think back to that experiment you did in high school where moving a copper wire through the open ends of a U-shaped magnet created an electric charge that could be measured by a galvanometer. Then perhaps your teacher got some big guy to turn a crank as fast as he could on a little mechanism that contained a coil of copper wire inside a magnetic field attached to a tiny Christmas tree light. Or perhaps it was a bicycle lamp and the gizmo was attached to a stationary bike peddled like crazy by the girl who someday would do a triathlon. The result was the same. As the Romans would say: *Fiat lux!* Let there be light!

Getting electricity from water works the same way, but instead of humans providing the power to turn the dynamo, moving water does. The scale is orders and orders of magnitude greater, of course. But inside the powerhouse in a hydroelectric project, what you're seeing is the magic of induced current in wire that is moving through a magnetic field, only on a huge scale.

While both water and electricity flowed from Hoover Dam as soon as it was finished in 1942, the irrigation component of Grand Coulee's mandate had to wait until after World War II. This is because the hydroelectricity wrested from the Columbia River there and at Bonneville, the dam closer to the Columbia's mouth, was needed to power the aluminum foundries, shipyards, and aircraft plants in Washington State and Oregon. What is maybe more important, power from these dams, as well as from those built in the Tennessee Valley in the 1930s, made possible the creation of the atom bomb.

Making plutonium at Hanford, beside the Columbia, and enriched uranium at Oakridge, in Tennessee, required immense quantities of electricity that, fortunately for the war effort if not for the following peace, was now available from the dams. Enough plutonium was produced at Hanford to make two bombs. The first, the test bomb, was exploded July 16, 1945, in New Mexico, and the second was dropped on Nagasaki on August 9. Enriched uranium from Oak Ridge went into the Little Boy bomb, the first atomic bomb used militarily and dropped on Hiroshima, August 6. Now thousands of litres of nuclear waste from making those bombs and the other nuclear weapons in the United States' arsenal are stored in concrete silos on the Hanford Nuclear Reservation. The location for the Hanford Project was not an accident: in addition to providing abundant electricity, the river furnished a constant supply of cold water to cool the reactors.[49] What might happen to people and animals who came in contact with the water downstream wasn't considered at the time because all efforts were focused on winning the war. The consequences and dangers remain, however, particularly in an area which as recently as six million years ago was inundated by massive lava flows, as witness the rocks that cover the hills above Grand Coulee. And, looked at one way, it's ironic that concrete was partly responsible for creating a danger that it alone was able to protect people from in the form of shelters, as we saw in chapter 3, "Fire."

Today there are fourteen large dams on the Columbia, eleven in the United States and three in Canada. Part of their mission is to control the rampages of the river: floods in 1948 that swept through the basin boosted pressure to proceed quickly in building more dams than Grand Coulee and Bonneville. The dams control floods but also even out the flow of water. The river runs much higher in spring and early summer as snow melts than it does during the nearly rainless summer. This fluctuation would make producing an even flow of electricity year round difficult. In addition, it presents problems for irrigation since the period of greatest water demand is when the river would naturally be at its lowest.

The series of dams has also transformed the lower Columbia into a series of what one critic calls "slack water pools," fundamentally changing the character of the river. Combined with dams and locks on the Snake River, they allow navigation as far as Lewiston, Idaho, 748 kilometres from the coast. By way of comparison, note that the St. Lawrence Seaway, which runs from Montreal to Lake Erie, is 595 kilometres long. The folks who planned and built the original dams in the 1930s might be surprised to see barge loads of wheat sailing through a wide concrete channel in the dry Palouse hills of eastern Washington, headed for China. In 2017, 50 percent of all U.S. wheat headed for foreign markets passed through the Columbia-Snake system as a whole.[50] The system's boosters emphasize that barge transport is safer and more environmentally friendly than moving that quantity of grain by either truck or train since one barge has the carrying capacity of 134 trucks or 35 jumbo hopper train cars.

Yes, the champions of this great wave of dam and lock construction might be surprised at that, but they would be astonished to learn that serious consideration is now being given to tearing down four of the dams on the lower Snake built in the 1960s and 1970s. At issue is the health of salmon and steelhead, those fish forgotten when Grand Coulee was built. Besides spawning in the upper Columbia, they once swam in the millions to spawn on the Snake and its tributaries. Five times—most recently in 2016—courts struck down plans formulated by federal agencies under endangered species legislation to preserve the fish by other means. The last time U.S. District Court Judge Michael Simon ruled that "breaching one or more of the dams" should be considered because of legal obligations under the Endangered Species Act and the National Environmental Policy Act Tribal Treaty.[51] There is no need to worry about the decrease in hydroelectricity if this were done, a study commissioned by proponents of getting rid of the dams argued in April 2018. The dams produce only 4 percent of the region's electricity, which could easily be replaced by solar and wind power augmented by energy storage and greater energy efficiency.[52] Indeed, hydroelectricity from the dams is now frequently priced higher than electricity from other renewable sources.[53]

The dams are still there, however. A three-year agreement reached in early 2019 by the federal agencies concerned, the states of Oregon and Washington, and the Nez Perce mandates increased water release by the dams that will supposedly improve conditions for the salmon. By 2021 a definitive study of the situation is supposed to settle the question once and for all. Until then, though, the jury is still out.[54]

Already, however, two dams built on the Elwa River on Washington's Olympic Peninsula have been removed in response to similar concerns. They were built in the 1920s to provide electricity for the lumber industry and were later used to power shipyards at Port Bremerton. They were put out of commission not only because of the damage they've done to the fish runs, but also because the aging dams would have required major repairs and the power needs in the area have shifted.[55]

It's true that in many places hydroelectricity might be replaced in part by other renewable energy sources. Today, power from Hoover Dam still flows to Southern California: during our visit to its powerhouse an additional turbine roared into action, prompting the guide to remark that it must be hot in Los Angeles because people were turning on their air conditioning. But the proportion of electricity used by Los Angeles that is produced by hydropower has dropped. Part of that is because the overall amount of electricity needed by the millions in the state is so large that the contribution from Hoover has become swamped by electricity from other sources. Gas- and coal-fired generators now produce 39 percent of its power, and solar and wind power are expected to produce more in the future.[56] Hydroelectricity production is also less reliable during droughts, which appear to be becoming more frequent as the climate changes.

Now the Los Angeles Department of Water and Power (LADWP) wants to turn the dam into a giant storage "battery" to complement solar and wind power installations. Water would be taken downstream from the dam and then pumped back through a huge concrete pipeline into the dam's reservoir, where it would generate electricity once again. Power to do the pumping would be generated by solar and wind at installations about thirty-two kilometres away. The pumping itself would be done when solar and wind were generating more power than could be used at the time. The US$3 billion project has many hurdles to overcome, but it would be much cheaper than an alternate storage system, the giant lithium-ion batteries being developed to absorb solar power and release it when needed, LADWP planners think.[57]

That's cutting-edge stuff, but in other places hydroelectricity, pure and simple, still comes with a great deal of glamour, and has for several decades. Take the James Bay and Manicougan projects in Quebec where I live now. The multidam systems begun in the 1960s and constructed in the 1970s are there only to produce electricity: they are much farther north than population centres so the flooding they caused wasn't considered a problem, and conventional agriculture in the surrounding territory was out of the question because of the rigors of the climate. No, what was front and centre was the idea that the Canadian province, home to eight million people of whom 70 percent had French as their mother tongue, could become a literal powerhouse, selling electricity elsewhere and using it cheaply at home to build a strong industrial base. The project was a long way from what was—and still is—called the Métropole of North America. It was the end point of an adventure that captured the imagination of a rising tide of people who wanted to be *maîtres chez nous*, masters in our own house.

When we arrived in this cold country from California, I didn't know anything about this, and I had no idea that hydro dams were imbued with the same kind of mythic power that the Columbia and Colorado projects had in the United States thirty and forty years before. What opened my eyes—and my ears—was a haunting ballad that was still getting radio airtime then, even though it had been released a couple of years previously. In "À la Manic," Quebec singer Georges Dor in effect reads a letter to his sweetheart from a construction worker on the Manicougan hydro project, about 900 kilometres from Montreal on the north shore of the St. Lawrence River. I listened to it over and over, puzzling over the tender and suggestive words, both because I was trying to learn French and because the longing in Dor's voice was compelling. Like Woody Guthrie singing about the Columbia, the song captured the spirit of the times.[58]

Begun in 1960, the Manic project includes eight dams on the Manicougan and Outarde rivers that cause water to back up into the crater left by a huge meteor or asteroid that crashed into the earth some 3.5 million years ago. Sometimes called "the Eye of Quebec," the crater originally was about 100 kilometres across, but has been somewhat reduced by erosion over the millennia. Before the hydro project, two large lakes existed inside the crater, but the rising water united them, leaving the high ground in the middle isolated as a large island.

In the end, more than 6,000 men worked on the complex dam structure. Cement was shipped from Montreal in a bulk carrier, making a journey like that cement from Port Daniel would make decades later, only in reverse. A short film by the National Film Board of Canada gives a glimpse of the conditions on the ship which made forty-seven round trips, only stopping in the dead of winter when the St. Lawrence became completely impassable because of ice. Aggregate was sourced on the site, and concrete on the centrepiece of the Manic project, the Daniel-Johnson dam, was poured 24/7.

Then Hydro Quebec, the public utility behind the great construction project, turned to another series of rivers in the north of the province, just to the east of James Bay. That was even bigger, and became a symbol of a place that knew how to do things. The project capitalized on one of the advantages of hydroelectricity: in principle, water that passes through a dam generating hydroelectricity can be used again and again, if more dams are built downstream. In the James

Bay project, a series of dams on the La Grande River turn seven sets of turbines.

But none of these dams would have made sense economically, had it not been for an innovation pioneered by Hydro Quebec: 735 kV transmission lines. The stepped-up carrying capacity allowed transmission over very long distances with proportionally less of the drop in power that comes from resistance to the current as it passes through the wires.[59] By way of comparison, the lines carrying electricity from Hoover Dam only carried 287 kV to begin with, while those at Grand Coulee after refurbishing now carry 500 kV. It is only in the twenty-first century that more powerful transmission lines have begun to be used, notably in China and India.[60]

The ecological fallout from these hydro projects, with their mammoth dams and the immense amount of electricity they produce, is in many respects less than that of dams elsewhere. The region is lightly settled, which has meant that massive population relocation wasn't necessary. The Cree of James Bay halted construction on that project until they got compensation for the use of their ancestral lands, and, even though caribou herds have been substantially reduced,[61] the Cree—also known as the Eeyou Istchee—seem to have suffered less than the Indigenous people on the Columbia, who saw the salmon fishery destroyed by Bonneville and Grand Coulee.[62] But they also say clearly that they don't want any more dams on their territory: that was the message they gave to Bill de Blasio, mayor of New York City, in July 2019 when he came to consult before negotiating a big electricity contract between Quebec and the Big Apple.[63]

Today, Hydro Quebec produces 36.36 gigawatts of hydropower. While it has a couple of dam projects under way at the moment, the great period of hydroelectric expansion seems past in Quebec.[64] And, as the controversy over the dams on the lower Snake suggests, the idea of building dams elsewhere in North America has few proponents, not least because most good dam sites have already been built upon. There are exceptions, though.

Millerton Lake, where I enjoyed swimming so much, is not very big in relation to the amount of water that flows down from the mountains in wet years, and excess is frequently released either into the river system or diverted to the Fresno urban area to recharge ground water. A waste, many elected officials and agricultural community spokesmen have argued for years. Their solution would be to build yet another dam near the far end of Millerton to hold water in wet times. However, the plan was effectively killed in 2018 when no funds were allocated to the project from a statewide fund designed to pay for more water management. Not really needed, the state legislators said. The subtext was that the state was now dealing with a climate regime where too little rather than too much water would be the rule and was the major problem to solve.[65]

Too little water may well be the cry of the future in many places, given the growing evidence that climate change will substantially disrupt rainfall patterns. Less water certainly is the problem on the Colorado now, in part because so much is taken from it during its passage to the sea, but also because of changes in rainfall. Lake Mead, the reservoir behind Hoover Dam, has only once had enough water for the excess to be let out into its spillways. That was back in 1983, and since then in most years the bathtub ring of salt-like sediment around the edge of the reservoir has grown and grown. By the time the river reaches what used to be its delta, the water is so saline and there is so little of it that Mexico has sued—and won—the right to have cleaner water in more abundance. In 2014 the water-replenishment scheme included in an international agreement went into effect, and the river reached the Gulf of California for the first time in decades. Even then, though, the situation was quite a change from that of the fall of 1922 when conservationist Aldo Leopold and his brother explored the delta. The visit was so remarkable that he never went back, for fear that he'd be disappointed. He wrote:

> When the sun peeped over the Sierra Madre, it slanted across a hundred miles of lovely desolation, a vast flat bowl of wilderness rimmed by jagged peaks. On the map the Delta was bisected by the river, but in fact the river was nowhere and everywhere, for he could not decide which of a hundred green lagoons offered the most pleasant and least

speedy path to the Gulf. So he traveled them all
and so did we. He divided and rejoined, he twisted
and turned, he meandered in awesome jungles, he
all but ran in circles, he dallied with lovely groves,
he got lost and was glad of it, and so were we.[66]

The waters of the delta were saline even then, however. Leopold says that finding good drinking water was the biggest challenge they faced. To test any water from a small well that they might dig or a promising creeklet they found, they seduced their dogs into trying it: if it were too brackish the animals wouldn't drink it. At the time, the salt in large part came in on the tides of the Sea of Cortez, the small arm of the Pacific that separates Baja California from mainland Mexico. The water that makes it to the delta today is much saltier, and also carries a heavy load of chemicals, pesticides, and other human pollution.

That is because irrigation has always had a definite downside. The Mesopotamians may have been the first to run into it 4,000 years ago when their irrigated fields began to produce less and less as salts carried naturally by their irrigation water built up in their fields. Many scholars suggest that their grand civilizations came to an end because they simply could no longer feed their people.[67]

The problem has its source in the way that water picks up salts and other minerals as it flows along. The amount and the kind influences the taste of the water. A spring in high mountains can produce smooth and sweet water as clear as glass. But by the time that water has run downhill, joining with other streams to become a river, the taste may change and the clarity disappear, even when there appears to be an abundance. John Steinbeck's Joads ran into that as they headed for the promised land of California:

The car pulled off the road and stopped, and
because others were there first, certain courtesies
were necessary. And the man, the leader of the
family, leaned from the car.
Can we pull up here an' sleep?

Why, sure, be proud to have you. What State you from?

Come all the way from Arkansas.

They's Arkansas people down that fourth tent.

That so?

And the great question, How's the water?

Well, she don't taste so good, but they's plenty.[68]

Water, especially when it is abundant amid penury, has a tremendous pull on the sojourners. When they cross the Colorado into the Golden State and are preparing to traverse the desert, one of the party decides he can't go on, but must stay with the water.

"Tom, I ain't a-gonna leave this here water . . . ."

"You're crazy," Tom said.

"Get myself a piece a line. I'll catch fish. Fella can't starve beside a nice river . . . I'll catch fish an' stuff, but I can't leave her. I can't."

He crawled back out of the willow cave. "You tell Ma, Tom."

He walked away . . . Tom started to follow, and then he stopped . . . . [And he] watched Noah growing smaller on the edge of the river, until he disappeared into the willows at last.[69]

The family runs into the vagaries of the river and the California winter a little later, when they are flooded out of a makeshift camp near Bakersfield during the winter rains, the kind of flood that the Central Valley Project was designed to master, the kind of flood that also flushed away the minerals and salts that had collected in the soil.

But after the dams were built, floods like that were a thing of the past, and the land was irrigated. What happens then is that plants suck up water but remove only some of the dissolved salts from it. What remains in the soil becomes concentrated as excess water is evaporated. Farmers now know this and to try to wash the salts away; they may apply even more water to carry the salts deeper into the soil below the plants' roots. The problem isn't really solved, however. The added water can percolate through the soil downwards to ground water, contaminating it, and, somewhat ironically, raising the water table. The result is plants that in effect are drowned by waterlogged soil in a dry country.

The damage was not immediately evident back in the days of the Mesopotamians, and it took a few decades for a similar sort of thing to develop in California's Central Valley. But it is a real problem now, made worse by the fact that as water passes through the soil today it picks up agricultural chemicals from fertilizers and pest control.

The Central Valley and the Colorado River Delta are not alone in suffering from the downside of irrigation. The worst example is probably what has happened to the Aral Sea in the southern part of what was the USSR. Until the 1960s two large rivers that rise in the mountains of Afghanistan, the Amu Darya and Syr Darya, flowed through desert country into the sea, which even then was the fourth largest saline lake in the world.[70] But, in order to "make the desert bloom," the rivers were dammed and their waters were diverted into irrigation projects, with an immediate effect on the level of the sea. By 1998 the water level was down by twenty metres, the water volume had decreased by four-fifths, and salt left by the evaporating water was blowing all over the region in huge dusty storms.

The wetlands of the Yellow River Delta are also in bad shape. Some of that is due to pollution from the petroleum industry that exploits the oil field lying underneath the delta, and to urban encroachment. But dams upstream have had an impact, just as dams on the Colorado have in the southwest United States. The number of days that the Yellow River Delta received no fresh water from the river grew dramatically: between 1972 and 1979, there were 86 no-flow days, but between 1990 and 1999 there were 902.[71] In 1997, the worst year, the river's water failed to reach the sea on 226 days. But since then, carefully calibrated releases from dams on the river have reduced that number to none.[72]

The Three Gorges Dam, currently the world's biggest, is not on the Yellow River but on the Yangtze, China's longest river, which flows across the vast country well south of the Yellow River. The Three

Gorges project had been discussed for decades before disastrous floods on the Yangtze's middle stretches in 1954 pushed the idea forward. It got another boost when Mao Tse-Tung took a much-heralded swim across the river in 1966, showing off his prowess—he was seventy-three at the time—and putting the spotlight on the river. A poem he wrote, called "Swimming," is said to address the wonders of the river, as well as the promise of what could be done with it:

> Sails move with the wind
> Tortoise and snake are still.
> Great plans are afoot:
> A bridge will fly to span the north and south,
> Turning a deep chasm into a thoroughfare;
> Walls of stone will stand upstream to the west
> To hold back Wushan's clouds and rain
> Till a smooth lake arises in the narrow gorges.
> The Mountain Goddess if she is still there
> Will marvel at a world so changed.[73]

But the Three Gorges Dam, which went into operation in 2000, is just the biggest monument to China's immense dam- and city-building spree.[74] Some 22,000 dams higher than fifteen metres have been constructed since 1950,[75] and the country's twelfth Five-Year Plan, set out in 2015, called for the construction of more than fifty large-scale hydropower plants. The list includes ones on rivers that aren't exclusively Chinese, including the Lancang/Mekong and Yarlung/Tsangpo/Bramaputra rivers, which flow through other countries during part of their run.[76] The latter dam projects, like dams on the Tigris and Euphrates in Syria and Turkey that affect river flow in Iraq and the ones on rivers that feed (or fed) the Aral Sea, have elicited storms of controversy because their downstream impacts are great. To build dams on the river that becomes the Mekong as it passes down the Indochinese Peninsula would be "devastating," according to a study commissioned by the Cambodian government and undertaken by the National Heritage Institute, a U.S.-based consultancy firm.[77] Because of disruption of the inland fishery, the food security of sixty million people in Cambodia, Laos, Thailand, and Vietnam would be affected. But the Chinese projects are not the only ones menacing these rivers. Laos is planning or has already built seventy hydropower plants on the Mekong and its tributaries, with most of the electricity slated to be exported to Thailand, Vietnam, and China itself. The dams themselves also can present immediate danger: one of them gave way in July 2018 during heavy monsoon rains. Thousands were displaced, raising many questions about whether these dams are being designed to withstand extreme weather events that appear to be becoming more and more common.[78]

Ironically, notes BBC reporter Phillip Ball, many of China's projects were undertaken for environmental reasons, that is, to provide "clean" hydroelectricity needed for China's burgeoning industry and its growing middle class.[79] Pollution from coal-fired generating plants is responsible for much of the choking smog that frequently covers vast areas of the vast country. Wade Shepard reports in his *Ghost Cities of China* that perpetual smog is the reason that solar power is not in the works for one new "eco-city," Tianfu, a satellite of Chengdu, China's fourth largest city, because not enough sun gets through to make solar power work.[80]

Note, too, that a move away from coal might also help preserve China's threatened water resources. Coal requires considerable water in order to be mined and prepared for market, while older coal-fired plants use much water for cooling equipment.[81]

China says it wants hydropower to contribute about 30 percent of all electricity generating capacity in the country by the end of 2020, ramping up to 380 gigawatts.[82] It appears to be well on track to do that: by 2018, the most recent year for which statistics are available, it had an installed capacity of 352 gigawatts. That represented 20 percent of all electricity produced in China, and, amazingly, 27 percent of the world's installed hydropower capacity.[83]

Water, of course, isn't valuable only for making electricity. More fundamentally it is an essential element for life as we know it, and the Chinese are now using concrete to try to redress serious water shortages. The problem, baldly put, is simple: China's northern plain is dry, but its southern provinces have a lot of water.[84] The region around the capital, Beijing, has had a chronic undersupply of water for decades, even though the location was chosen because of its abundant springs and streams back in the eleventh century BCE when the city was founded. That has changed dramatically: by 2014 the Wanquan River, whose name means "ten thousand springs," had been reduced in Beijing to an urban channel that had been concreted so that what water it carried wouldn't soak into the ground, according to local environmentalists.[85] The amount of water available per Beijing resident then was about 120 cubic metres per person a year, well below the 500 cubic metres the UN deems a situation of "scarcity."

Furthermore, the North China Plain, on which Beijing sits and through which the Yellow River runs, produces a major part of the country's food. Watering those fields was a major goal of a government where engineering projects had high priority: former President Hu Jintao, in office from 2003 to 2013, was himself a water engineer.[86]

The solution they came up with is simple to describe, but much harder to put into effect: move water from the south to the north. As Mao wrote, "A bridge will fly to span the north and south, / Turning a deep chasm into a thoroughfare."

Southern California girl that I once was, I have a gut reaction favourable to the idea of getting water for ordinary folk to drink, wash, and otherwise use. That house we lived in—built in 1947, you'll remember—would never have been constructed had it not been for water from Hoover Dam and the Colorado River. The streets in our neighbourhood had been platted sometime before World War I, but it was clear by the 1930s that Southern California would never be developed unless other sources of water were found. The rather erratic rivers that flowed through the region and the aquifers that for a time fed wells simply couldn't provide enough. Even water from the Owens Valley project that supplied the Los Angeles basin beginning in 1913 fell short of demand. By 1940 and World War II, providing water for the thousands and thousands of people who flocked to work in war industries had become a priority. (Many of them, by the way, were people like the Joads who'd been on the coast long enough to figure out how things worked.) All that concrete poured, all those millions of megalitres of water impounded behind those New Deal dams helped the war effort just as electricity from the Columbia and in the Tennessee Valley did.

But only in 1947 did water begin flowing to San Diego from the Colorado River through a pipeline jointly built by the U.S. Navy, the Bureau of Reclamation, and the San Diego County Water Authority.[87] Another pipeline was built in 1952, the year we moved to San Diego from Washington State. It was just another in a long line of water movement projects that go back to antiquity.

MOVING THE WATERS

The Romans, of course, were aces at building aqueducts, and their concrete structures can be seen wherever they settled. One of the most impressive is the one set on a bridge supported by three tiers of arches that span a valley near Mérida in Spain. It carries water from the Proserpina dam, which we mentioned before as being one of the oldest still-functional concrete dams.

Closer to home and more recently, the Catskill Aqueduct that supplies water to New York City was cut through mountains north of the city in 1915.[88] A 160-kilometre trench was first dug and blasted through rock, and then concrete was poured into the bottom and sides of the tunnel. Once that was set, circular formwork allowed concrete to be poured on top, making a huge pipe through which the water would flow. What remained of the trench was filled with earth, and—voilà!—water could run from reservoirs in the Catskills downhill all the way—and it still does.

The great dam and canal projects in the American west are much larger in scale. Three carry water more than twice as far as does the Catskill Aqueduct: the Colorado River Aqueduct, 389 kilometres; the Central Arizona Project, 541 kilometres; and the combined 1,128 kilometres reservoir, pipeline, and canal system of the California State Water Project.[89]

But the North American solutions are just a drop in the bucket compared to China's South North Water Diversion Project or SNWD. Indeed, practically no water is transferred from the south to the north on this continent, except for that contained in the millions of tons of fruits and vegetables grown in irrigated fields of the Southwest and California. That sort of water can move a long way, as witness an outbreak during the winter of 2018 of bacterial contamination from romaine lettuce, which contains about 95 percent water.[90] Sickness from a specific strain of E. coli bacteria first showed up when eight inmates at an Alaska prison fell ill after eating the lettuce. Before the exact source of the outbreak was traced to a concreted irrigation canal in Arizona, more than 200 people were sickened and five died, two in Minnesota and one each in Arkansas, California, and New York.[91]

But when it comes to water transfers, the only thing in North America that might remotely resemble China's SNWD is a version of an idea floated in the 1970s by Quebec's premier of the day, Robert Bourassa, who suggested that water from the great rivers of Quebec's hinterland might be shipped south to slake the thirst of Americans. That possibility was brought up again in 2008 by a Montreal-based think tank, but so far nothing has come of the idea.[92] The same holds true for semi-serious suggestions that would send water south from the great rivers of western Canada—the Fraser or the Mackenzie perhaps—or the Great Lakes. As David Owen notes, the water might be "wheeled" from one state to another, rather than piping it or transporting it in canals. The Great Lakes, remember, drain toward the Atlantic via the St. Lawrence, but a shift from the westernmost lakes in the other direction could be engineered. Precedent exists: sewage from Chicago, on Lake Michigan, has been pumped over the height of the land into the Mississippi watershed since the turn of the twentieth century. Clean water from one of the Great Lakes could be diverted into the Illinois River or the Missouri River, with another diversion into the Kansas and ultimately into the Colorado, where it would pass through Hoover Dam and straggle onward to the Gulf of California.[93]

But China's SNWD has already accomplished far more than these far-fetched scenarios would. First proposed in the 1950s, the SNWD plan is designed to move massive amounts of water along three corridors in order to provide enough water for Beijing to function, the North China Plain to grow food, and Chinese industry to prosper. The eastern corridor, opened in 2013, more or less follows the ancient Grand Canal, construction of which goes back to the fifth century BCE, but extends beyond it to Tianjin on the Bohai Sea. The middle corridor, opened in 2014, carries water from a reservoir on the Han River, a tributary of the Yangtze, as far as Beijing, with a spur to Tianjin also. When all three corridors are finished—and the third, most westerly one will not be built for at least thirty years—water will run through 4,350 kilometres of canals, pipes, and reservoirs. Already, Beijing is getting 70 percent of its water from the project.[94]

What has been built features concrete-lined canals that are cordoned off from the farms on the land they pass through. Both of the existing aqueducts dive under the Yangtze in tunnels, and both have required wholesale movement of population to make room for both the right of way and the reservoirs along the route: as many as 350,000 people are estimated to have been displaced. The third corridor aqueduct would present problems of a different order: it would be in the headwaters of the rivers and would require bringing the water over 3,000-metre-high passes. In addition, as mentioned earlier, it would change the amount of water flowing into other rivers that only begin in China and then run through other countries, all of which rely on them for water, fishing, and industry.

Will this enormous undertaking solve China's problem of water imbalance? Critics both inside and outside of China have their doubts. Making water more expensive would encourage better use of what water there is, some say. It is necessary to attack both the demand and the supply side of the water equation, they add.

Maybe the story of Fresno, just south of Friant Dam, might show the way. Despite how close it is, the city never took water from the dam because, until about ten years ago, groundwater was present in sufficient quantity so that enough could be pumped from the aquifer to satisfy the area's household and industrial needs. Nevertheless, in the 1990s a certain amount of "recharging" of underground water was begun by creating ponds to store rainwater or excess water from Millerton when the lake was nearly full. The idea was that water would percolate back into the soil, and to some extent replenish the underground supply. The effort was begun not a moment too soon because groundwater, which once was just three metres below the surface, retreated during the drought of 2011–2016 to ten times deeper. The water district now has the right to draw some water from Friant but can't count on it for supplying the urban area's half million population. Agriculture is higher in the arcane ranking of water use rights than is consumer use. So, to ensure an adequate supply, Fresno cracked down on a wide variety of water uses, cutting the city's consumption by more than 30 percent.[95]

The city is not alone: California-wide water consumption restrictions went into effect in 2020. They start out at 208 litres of water for indoor use per person and will decrease to 189 litres per day by 2030, and there will be penalties for water districts that don't comply.[96] The restrictions are completely doable, all agree: following emergency drought measures between 2014 and 2018, water consumption across California dropped to below what had been consumed in the benchmark year of 2013.[97] This spectacular success, however, must be greeted with caution. The temptation will be to allow more development because there appears to be more water. As David Owen put it in his book about water resources and the Colorado River, "Making people more efficient at using water isn't a gain for the environment if the gains are reinvested in sprawl."[98]

Over the centuries the Yellow River has dropped its load of silt as it reached the arm of the Pacific called the Yellow Sea. At the moment, as we've seen, it enters tidewater to the north of the Shandong Peninsula and sometimes the flow is greatly reduced. Several other rivers that have meandered across the North China Plain also end here in the Gulf of Bohai, bringing with them large amounts of sediment. The resulting shoreline around the gulf should be rich with fish, birds, and other sorts of life thriving in the tide flats and beaches of the rivers' deltas. Dam construction upstream, the water transfer projects, and the tremendous growth of inland cities have damaged that environmental bounty. Nevertheless, by July 2019 enough remained to convince the UNESCO World Heritage Committee to name the migratory bird sanctuary on the Bohai Gulf a World Heritage site. It is, says UNESCO, "one of the most diverse and fertile coasts in the world, a key habitat for the migratory birds on the East Asian-Australasian Flyway. Besides, habitats formed by thousands of years of human activities, including rice fields and salt works, also provide stopover sites for migratory birds in certain periods of the year."[99]

But there are hydrocarbon reserves underneath the gulf, and the shipping channels in the strait leading into it are among the busiest in the world, so what will happen in the future there is an unanswered question. The port at Qinhuangdao, the city at the top of Gulf of Bohai, is located at 39.9 degrees north latitude, or slightly farther south than New York City, which makes it the most northerly of China's ice-free ports. Coal is abundant in the hinterland to the north, and port facilities see more than 250 million tons of coal pass through its installations every year.[100] But the port is only one part of the city of about three million, which is divided into three major administrative districts: the city proper centred on the harbour; the northern section where the Great Wall of China begins; and Beidaihe, the southern section which has been a summer retreat for China's elite for more than a century, no matter who has been in power.

Since the beginning of the Communist reign, Beidaihe's golden beaches are where party leaders have met each summer to plan the party's and the country's future. The resort town is now a two-hour train ride from the heat and air pollution of Beijing, but even before that, Mao reportedly loved to swim in the waters of the bay. It is also here that concrete's powerful but ambivalent influence comes sharply into focus.

In 2010 local and national governments adopted a comprehensive plan for Beidaihe that might be a model for any region that wants to safeguard a coastal region of exceptional beauty and environmental richness. Sasaki and Associates, an international urban design firm begun in Japan and now with offices in the United States and China, drew up guidelines for integrating new construction with wetlands preservation. Presented in 2010, for several years it had pride of place on the firm's website, showing drawings of birdwatchers and people rowing on tidal channels, as well as a raft of attractive cluster developments oriented toward the water.[101] Nearly ten years after the report was finished, it's hard to tell how much has been realized, although several signature buildings have been constructed a bit farther south on the beach itself. They include a library and a church, both designed by Vector Architects, the brainchild of Beijing architect Gong Dong. Both are also beautiful examples of careful construction with concrete and have received rave reviews in architectural journals.[102] How well they'll stand up to rising sea levels and high-tide waves is unclear, however, even though Gong Dong says that the library was built after careful analysis of predictive climate models.

Someone who doesn't know the ins and outs of the area might be tempted to call it all just show; I've never been there, I can't say for sure. But there is one nearby development that hints at what might be done to redeem a world defaced by constructions made possible by modern concrete: Moshe Safdie's Habitat Golden Dream Bay community. It is his most ambitious plan to provide, he says, affordable housing and "for everyone a garden." While it's not clear what "affordable" means in this context—Qinhuangdao is classified by the Chinese government as a "third tier" Chinese city and housing is not as expensive as it is in first tier cities like Beijing or Shanghai—the complex appears to have been enthusiastically welcomed.[103] Its first phase, completed in spring 2018, included 2,500 residential units, nearly 18,300 square metres of retail space, and villas along the beachfront. The second phase of construction will add three towers and two adjacent stepped buildings, as well as villas.[104] Across the road and on the other side of a grassy common lie the beach and the blue waters of the Gulf of Bohai. The view from the upper storeys of Golden Dream Bay must be even more spectacular than those of the original Habitat in Montreal, only here, instead of a river, the water is the sea.

But change is in the wind. Until now the sea has been extremely important for Qinhuangdao. Nevertheless, while it is still one of the busiest ports in China, it appears to be left out of the Belt and Road Initiative (BRI), the challenge that China has set itself now that the goal of urbanizing its population is nearly accomplished. The multi-billion-dollar plan is aimed in part at bringing economic development to the western part of China. Building it will require a lot of concrete, as well as steel, glass, and other materials, a surfeit of which the country seems headed for. Roads and high-speed railways from the dry western regions will cut south to centres on the Indochinese Peninsula and across the mountains to India and Pakistan: this will be the Silk Road Economic Belt. Ports from Hanoi to Nairobi to Venice will be linked in what the Chinese are calling the 21st Century Maritime Silk Road. The project is vast and so will be Chinese investment, not only in China but in many other countries. An example is the port installations in Colombo in the island-nation of Sri Lanka. Chinese interests financed and built the project, but the Sri Lankese have been unable to pay back the investment. The result? Default by the locals that gave China control of important installations not far from India, its biggest rival in Asia. The port, says *The New York Times*, is "a strategic foothold along a critical commercial and military waterway."[105]

As for Qinhuangdao, with the advent of the BRI it may lose much of its status as a hub of coal shipping: there's a project in the works to move most of that business south to Caofeidian, some 240 kilometres away and handier to the expanding Maritime Silk Road routes.

So what Qinhuangdao and its gold beaches should do and will do is concentrate on tourism, say observers.[106]

One of the current attractions is located just north of the city, along the coastline beyond where the coal is loaded now. The Dragon Head is the beginning of the Great Wall of China. Extending twenty-three metres into the waters of the Bohai Sea, the structure—which looks something like a dragon stretched out into the water, lapping up the element—is not the original one that guarded the Chinese from northern invaders four or five hundred years ago. That one fell to disrepair and had all but disappeared by the 1980s when it was rebuilt with a concrete core and bricks for scales. The re-creation is currently a top tourist attraction: one Trip Advisor comment says it's the nicest place on the Great Wall to visit in the vicinity of Qinhuangdao. And certainly the restoration work appears more tastefully done than another repair job on the wall a bit farther west, where the local authorities laid down concrete along the top.[107] It's much easier to walk on now, the mayor explained when outsiders complained that the repairs made the wall look like a "skateboard track." But as British expat Andrew Killeen wrote shortly after visiting two other sections of the wall with his family: "The conundrum remains: how one of China's greatest cultural symbols can be kept authentic without becoming dilapidated and overgrown; how tourism can best be managed and generate the funds for the Wall's upkeep, without turning it into a theme park."[108]

But it keeps tourists happy, and that would seem to be the future of this growing, formerly quiet resort that, apparently, Mao the poet loved. He wrote in 1954:

> Beidaihe
> Summer
> A rainstorm sweeps down on this northern land,
> White breakers leap to the sky.
> No fishing boats of Qinhuangdao
> Are seen on the boundless ocean.
> Where are they gone?
> Nearly two thousand years ago
> Wielding his whip, the Emperor Wu of Wei
> Rode eastward to Chiehshih; his poem survives.
> Today the autumn wind still sighs,
> But the world has changed![109]

Yes, the world has changed, and with it the quality of the water of the rivers and the seas, transformed in part by the concrete of which it is such an important element.

CHAPTER 5
AIR

INSUBSTANTIAL AS A CLOUD?

The last of the elements to consider here is air, which at first glance appears to have little to do with concrete since concrete is such a solid substance. But there is air in most modern concrete—up to 5 to 7 percent in fact—and the physical presence of air in the material is only one way this Roman element is connected to the concrete that has built our world.

Air in concrete—called entrained air—consists of billions of tiny bubbles whose benefits were discovered more or less accidentally, another example of concrete guru Rick Scezcy's adage: "If you got a pile of crap, someone is going to say, let's try it in concrete and see what happens."[1] That's not exactly what transpired, but it demonstrates equally well just how inexact a science making concrete has been and continues to be.

In the 1930s a number of state public works departments were looking for ways to improve the durability of concrete, particularly its resistance to scaling when exposed to road de-icing salt. The New York State Highway Department found that mixing 15 percent "natural" cement—that is, the kind of cement made from calcinating rock containing limestone and clay which we talked about back in the chapter "Earth"—with 85 percent Portland cement resulted in a concrete that was much more durable than concrete made with regular Portland cement alone. No one was quite sure why until it was noticed that of the two kinds of natural cement used, the one that performed best contained a small amount of beef tallow used as a grinding aid. Then it was found that the best kind of Portland cement had been contaminated with a small quantity of oil or grease that had leaked from the bearings that crushed the clinker into powder. More experiments followed, and by the middle of the 1940s it was discovered that the additives created something like soap lather when the concrete ingredients were mixed together. As the concrete set, the bubbles remained. You can get an idea of what happens on a macro scale if you blow a soap bubble in extremely cold weather. Below about -15 degrees C the bubble's surface freezes in a lattice-like pattern rather like the lattice that forms as concrete sets.[2] The bubbles in concrete are microscopic, though, and what they do is provide places for water to expand when it freezes: no bubbles and the frozen water could literally split the concrete apart.[3]

Romans also incorporated air in their concrete, but in a completely different fashion and for different reasons. Freeze and thaw cycles aren't all that common around the Mediterranean, after all, but engineering domes and arches was a constant in their construction projects. Empty, sometimes cracked, upside-down amphora—those curvy terracotta vessels used to store such things as wine and olive oil—are found in many of them, along with more conventional aggregate. The supposition is that including them reduced the weight of the dome or arch, and so helped maintain its structural integrity. Similarly, lightweight volcanic tuff shows up as part of the *caementai*, the rubble onto which Roman cement was slathered. While tuff might have had some of the properties of the volcanic sand that is the secret of Roman concrete, it also is many times lighter than ordinary rock because it is honeycombed with air spaces. A brilliant example of this use is found in the gorgeous dome of the Pantheon in Rome.[4]

Concrete has two other important connections with air. The first is as tangible, physical, and measurable as the substance itself: the effects of making and using it on the planet's atmosphere. The second is metaphoric, and follows from the fact that the human spirit sometimes seems as insubstantial as air: it is what concrete, that wonderful, surprising material, can call forth from our souls.

But first concrete and the atmosphere.

CARBON DIOXIDE

As I said near the beginning, the air when I visited the McInnis Cement plant at Port Daniel was sparkling. Days of rain had cleared out the atmosphere, and blue skies with fluffy clouds made the view across the Anse McInnis picture-perfect. But this was just before production started, and shake-down trials had been temporarily stopped because of a glitch in the bearings of the rotary kiln. When the plant was up and running, scrubbers would remove most of the particulate matter from the gases that would go up the smoke stacks, I'd been assured. Visible air pollution would be minimal.

However, the air would then contain a much higher percentage of an invisible gas than it did before production began. The gas, carbon dioxide, is what we've come to know as the poster kid for greenhouse gases, for climate change, for melting ice caps, for rising sea levels. McInnis says it is working hard at finding ways to cut down on CO_2: the plant is far more efficient than the competition because of its up-to-date kiln, they assert, adding that the plant will be the least polluting per ton of cement of any around. Yet by its own estimation, McInnis will be the single biggest CO_2 emitter in Quebec, while, worldwide, cement production is responsible for between 4 to 6 percent of all CO_2 emissions. Smokestack emissions don't tell the whole story, though: concrete's CO_2 footprint is larger than that and continues long after cement has become concrete.

Two elements of the process of making cement produce CO_2 directly. First, the temperatures needed to reduce limestone or other similar rock to its basic components are extremely high, requiring burning great quantities of fuel. Second, the chemical process itself liberates much CO_2, as we saw in the initial chapters. After all, what's happening is that calcium carbonate is becoming calcium hydrate, and what's left over is CO_2.

Thirdly, and perhaps most importantly, the way of life that concrete has made possible is responsible for an immense amount of greenhouse gases. This last contribution is somewhat difficult to quantify but we'll make an attempt because otherwise we really won't have an appreciation of what concrete has done to our world.

Let's look at air as a physical substance first. Of the four elements that the Romans thought made up the world, air is perhaps the hardest one for a person of our times to get his or her mind around. Air surrounds us, much of the time it is transparent, it is easy to forget that it is there. But for the ancients it was as elemental as earth and water, and we know that it is as essential to life as they are, perhaps even more so.

On average, the air we breathe is made up mostly of nitrogen, some oxygen, some hydrogen in combination with oxygen (water), and soupçons of many of the elements recognized by modern science. The usual proportions run something like nitrogen, 78 percent; oxygen, about 21 percent; argon, almost 1 percent; as well as carbon dioxide, and small amounts of other gases. Air also contains a variable amount of water vapour, around 1 percent at sea level, and 0.4 percent over the entire atmosphere.[5]

The main components of air—nitrogen, oxygen, and argon—are like open doors, allowing light and energy in and out of the earth's atmosphere without impediment. That means that if they were the only components of air, energy from the sun would pass right through to warm the earth, and then be reflected back out into space almost as if the air were not there. But carbon dioxide, water vapour, and other gases like methane that are present in lower quantities absorb part of the heat radiated back by the earth. This creates a sort of gaseous blanket holding the heat in, the way that glass in a greenhouse does, which is where the name comes from. The more greenhouse gases the atmosphere contains, the greater the heating effect will be. Carbon dioxide is used as a proxy for the increase in all greenhouse gases, since when it goes up, it's almost certain that methane and the other greenhouse gases increase also. Therefore, the progression of CO_2 rates in the atmosphere is a handy shortcut to understanding a larger problem.

In the past, there have been times when the atmosphere contained much more carbon dioxide than it does now, but that was hundreds of millions of years ago, when the earth was much warmer and much lonelier.[6] It was more like our sister planets Mars and Venus, which now have atmospheres full of carbon dioxide: Mars has about thirty times more than Earth does, and Venus has a whopping 300,000 times more.[7]

What made the difference here is life, or rather the kind of life that has developed on Earth. Most of Earth's wealth of carbon (and there's no good reason to believe that all three planets started off with greatly different amounts of carbon) has become an integral and important part of plants, plant residues (including fossil fuels), and the skeletons of organisms such as those tiny ones who live in oceans, fall to the bottom, and over time became rocks like the limestone that we make into cement. Thus, slowly, slowly, carbon in the form of carbon dioxide was removed from the atmosphere and more or less locked up. Locked up, that is, until we started liberating it in massive quantities in the nineteenth century by burning fossil fuels and deforesting wide regions of the planet. Up until that point there had been a rough equilibrium for several million years between the carbon dioxide produced by natural processes such as forest fires and CO_2 given off by plants when they are not photosynthesizing and the CO_2 absorbed by growing plants and shell-forming animals.

As it happened, my trip to McInnis Cement came a few days after the day when the highest level of CO_2 was recorded for that year. On April 26, 2017, the climatological research station on the Hawaiian volcano of Mauna Loa recorded 412.63 parts per million (ppm) of CO_2, the highest of the year in the northern hemisphere. It was followed by 411.27 ppm, recorded on May 15, 2017.[8] The observatory there is part of the Global Greenhouse Gas Reference Network, which since 1958 has been measuring gases that are strongly linked to climate change—CO_2, methane (CH_4), and nitrous oxide (N_2O.) Carbon monoxide (CO) is also measured because it is thought to be a strong indicator of ordinary, visible air pollution. Observations are taken hourly on the volcano, as well as at another northern hemisphere site at Barrow, Alaska, and two in the southern hemisphere, at the south pole and on the Pacific island of Samoa.

CO_2 measurements vary with the seasons. They rise during the winter when many plants stop growing, and decrease in spring and summer when plants begin growing again and take up CO_2 as they photosynthesize.[9] But independent of this annual cycle is the recent continuous increase in greenhouse gases, particularly CO_2, due to the burning of fossil fuels. If you graph the ups and downs, what you get is a regular jagged pattern that is rising toward levels that most scientists say are going to spell deep trouble for civilization as we know it. The climb continues: in 2019 the high came on May 11, when 415.26 ppm were recorded.[10] What's in store are melting ice caps, more extreme weather, and rising sea levels, according to the men and women who have been studying what we've been doing to the atmosphere over the last 150 years. Warmer air means that more water will be evaporated from the oceans and will fall according to patterns that are only now being to be understood: the increase in "exceptional" weather events, however, is one of the few things that all models of climate change predict.

While there still are doubters, the evidence amassed during the last three decades has convinced many countries to try to turn back that trend through international agreements. The first attempt, the Kyoto Protocol, set legally binding emission reduction targets for developed countries, although the United States never signed it and Canada pulled out of it. Nor did Russia, Japan, and New Zealand sign the protocol's second stage, the Doha amendment, which would see participating countries reduce emissions by at least 18 percent below 1990 levels. A new agreement reached in Paris in 2015 aimed to cut emissions enough to keep the global average temperature to "well below 2 C above pre-industrial levels" and to "pursue efforts to limit the temperature increase to 1.5 F, recognizing that this would significantly reduce the risks and impacts of climate change . . . ."[11] Developing nations—which more or less got a pass in the Kyoto Protocol—were included in this pact, but the United States pulled out of it after Donald Trump became president, so just how effective it will be is clearly—pun intended—up in the air.

At the moment two major strategies exist whose aim is controlling CO_2 emissions. Both make individual businesses who emit CO_2 pay a price for doing so. One is a tax charged directly to emitters based on how many tons of CO_2 they emit. The other is called "cap and trade." Under it, a government sets a limit on CO_2 pollution from industry and sets up pollution quotas that companies are given or buy through auctions. The "cap" is reduced each year, and companies must bring down their emissions, buy quota credits from other companies who have successfully reduced their own emissions, or pay fines. The price of quotas is determined by the market, which makes the method more palatable to some entrepreneurs and governments. Typically, the money raised by the fines and sales of credits goes into a fund earmarked for green initiatives. The European Union has had such a system since 2005 and has seen CO_2 emissions fall by 8 percent: the goal is that the sectors covered by the system will be 21 percent lower by 2020.[12]

California and Quebec both bought into the idea too.[13] The money Quebec has collected—c\$3 billion through 2018—goes to green initiatives that include aid to businesses and industries cutting their own greenhouse gas emissions, installation of electric car recharging posts at work places, and transformation of compost and green waste to methane.[14]

Ontario, which was going to join forces with California and Quebec, abruptly pulled out of the scheme in 2018 after the election of a Progressive Conservative government under Doug Ford, who campaigned on killing it. In contrast, nine New England and mid-Atlantic states—Connecticut, Delaware, Maine, Maryland, Massachusetts, New Hampshire, New York, Rhode Island, and Vermont—have set up a similar plan for emissions from the power industry, but not the cement industry.[15]

Another way to force a reduction is to place a tax on CO_2 per ton. The approach has been tried in British Columbia since 2008, rather successfully, although it should be noted that the Cement Association of Canada has complained about competition from cement produced in the United States and China where similar taxes haven't been levied.[16] The carbon tax in BC began at c\$10 per ton, and rose by c\$5 a year: in 2021 it will top out at c\$50 a ton. The money collected is largely returned to citizens in the form of reduced sales and income taxes. That element probably accounts for the generally favourable attitude to the tax among people in BC.[17]

When carbon taxes—either cap and trade or a straight tax—are not what the economists call "revenue neutral," opposition to them can skyrocket. Protesters took to the streets in France when the government of Emmanuel Macron imposed a similar tax in 2018, the proceeds of which were supposed to pay down the government debt, not reimburse taxpayers. The result was months of militant protest—the *gilets jaunes* movement. In the end, the Macron government responded by cutting a proposed fuel tax increase, but remained tough on the carbon tax itself.[18]

Other governments have buckled to pressure. Six years before the French plan, Australia launched a cap and trade program which effectively set the price at c$23 a ton. It stirred up fierce opposition, playing a big role in bringing down the government that brought it in. A new, more conservative government elected in 2013 promptly abolished the tax, and business continued as usual:[19] Australia is now fifty-seventh among fifty-seven countries when it comes to climate-change action and is the world's largest exporter of coal. Australian novelist Richard Flanagan and other climate change advocates say there's a straight line between that attitude and the fires that raged there at the end of 2019 and the beginning of 2020. Writing in *The New York Times* on January 3, 2020—a day when hundreds of Australians waited on beaches to be rescued from advancing flames—Flanagan quoted Mikhail Gorbachev, the USSR's last leader, and held out the hope that, just as the horrendous nuclear accident at Chernobyl marked the beginning of the end of the USSR, so the fires of 2020 might prove to the "Chernobyl of climate crisis."[20]

That remains to be seen, but it's true that carbon taxes can be accepted when handled properly. Portugal instituted its own in 2015 with rather little brouhaha: the burden on the ordinary consumer was offset by reductions in the tax on gasoline.[21] The result was that the country had the best CO_2 reduction record in the European Union in 2018.[22]

In 2019 the Canadian federal government under Prime Minister Justin Trudeau began to impose its own carbon tax in provinces which had not set up their own carbon pricing schemes. The levy started out at c$20 a ton and will rise to c$50 a ton by 2022, with much of the money raised returning directly to citizens through what's called the Climate Action Incentive.[23] The tax is not charged on all emissions, but on the amount of CO_2 a plant produces in excess of a benchmark, which varies among industries. When the initiative was announced, the benchmark was set at 70 percent of an industry's average emissions performance, but subsequent changes mean that cement plants will have to pay a tax on, at most, 10 percent of CO_2 produced in excess of the industry benchmark.[24]

Nevertheless, the measure has elicited strong opposition in five Canadian provinces—New Brunswick, Ontario, Manitoba, Saskatchewan, and Alberta—who are contesting the right of the federal government to impose such a tax. Quebec is joining them in the legal battle, not because it is opposed to carbon taxes but because it doesn't like the idea of the provinces being forced by the federal government to do much of anything.[25]

That's an internal Canadian problem, though. The United States is faced with other, larger ones, in part stemming from a federal government under President Donald Trump which doesn't recognize the existence of greenhouse gas problems. This is despite the advice of a group headed by market-oriented lawmakers who might be thought to be against taxes on principle but who are championing carbon taxes. Americans for Carbon Dividends, supported by Conservation International, some fossil fuel giants, and companies in renewable and nuclear energy and consumer goods, is advocating a carbon tax plan "based on the conservative principles of free markets and limited government, and offer[ing] the most popular, equitable and politically viable climate solution," according to their mission statement.[26] How much influence on decision makers in the United States and elsewhere they will have remains to be seen, but they certainly broaden the range of players in the game.

What is equally unclear, however, is just how high carbon taxes must be to prompt a significant enough difference in our behaviour to save us from a climate change apocalypse.[27] William Nordhaus, who has been studying carbon taxing for four decades and who jointly won the 2018 Nobel Prize for Economics with Paul M. Romer, thinks perhaps the best way to avert complete climatic disaster would be to get countries to agree to a worldwide target carbon tax of perhaps US$23 a ton, with abstaining countries being punished perhaps by special tariffs for not complying. (Nordhaus thinks the BC model is one that should be widely adopted, by the way.)[28]

Pie in the sky? Probably. And, sadly, there have been moments in the past when it appeared possible to turn back climate change. Most notable was the decade of the 1990s when, as Nathaniel Rich outlined in his long article in *The New York Times*, "Losing Earth," the United States, Canada, and most European nations were near a tough, comprehensive plan to cut back CO_2 and other greenhouse gases.[29] That agreement fell apart, and we're living with the consequences. Another big unknown is how governments in the developing world will act: China has begun some carbon taxing initiatives in seven regions, but as we'll see, it's far from clear what that will accomplish.

When CO_2 measurements first were made at Mauna Loa, the carbon dioxide level stood at 316 ppm, which was higher than the estimated level before the Industrial Revolution of 280 ppm.[30] Between 1950 and 2017, emissions from humans jumped from 5 billion tons of CO_2 per year to 35 million tons, with the resulting increase in atmospheric CO_2 that we're concerned about.[31] The growth slowed between 2014 and 2016, but in 2017, the year I visited McInnis Cement, the trend went in the other direction, increasing by 1.4 percent to 32.5 gigatons, a record high.[32] The following year carbon emissions went even higher, even though emissions from countries in the European Union remained steady, although they'd been trending downward, as most member nations pursued policies in line with the Paris Accords on climate change.[33] Note that some think this is a bit deceptive because much of the goods consumed in the EU are actually now manufactured in countries like China, where CO_2 emissions continue to rise. Indeed, China was the single most important driver of the increase in global emissions: in 2018 its fossil fuel consumption rose, even though an increasing part of the country's energy needs were being met by the hydroelectric projects that have recently come on line.[34] In part that's because of lower than usual rainfall, which made hydroelectricity projects less productive than planned and which in turn meant more reliance on coal for producing electricity.

CO_2 emissions in India have been growing at a slower rate: the increase dropped during the last decade from 6 percent a year to 2 percent, in part because there was a 6 percent reduction in cement production. That last statistic is instructive. As said before, the cement industry in recent years has directly accounted for between 4 and 6 percent of all CO_2 emissions, while transportation has accounted for about 15 percent.[35] A decrease in emissions from cement-making could have major impacts on the effort to keep the proportion of CO_2 in the atmosphere from climbing higher. How to do that is a thorny question, though.

Three ways which might be coupled with carbon taxes suggest themselves. The first is efficiency in cement production. A second is changing the amount of cement used to make concrete. And a third, but perhaps the hardest to accomplish, is a radical move away from using concrete. This might come because of decisions made worldwide to work toward a low-growth economy, or it could happen because the way of life we are collectively living collapses because of drastic climate change or massive civil upheaval.[36]insert new footnote: The Covid-19 pandemic could turn out to be the catalyst for such a change. This book was being finalized for publication during the first months of the pandemic, when the long-term environmental and economic effects of the virus were unclear. Something of the latter happened in the 1990s, it seems, when the collapse of the Soviet Union led to wholesale abandonment of collective farms as well as drastic decreases in economic activity in Eastern bloc countries. This double whammy both lowered the amount of CO_2 going into the atmosphere and also removed some of what was already there by locking up some CO_2 in the plants that flourished in the fallow fields.[37]

Careful analysis of cement production data shows that such dips in emissions from cement production were particularly marked in the Baltic states of Estonia, Latvia, and Lithuania, and in Ukraine and Russia. Similarly, after the financial crisis of 2008 cement production dropped abruptly in several countries: Iceland, Greece, and Ireland, which had enjoyed spectacular growth in construction in the early 2000s, were hard hit, as was the United States.[38]

There is another menace looming that lies beyond the scope of this book, but which nevertheless should be mentioned: the prospect of a quick transition to a low-carbon economy through quickly dropping prices for renewable energy technologies. In a paper published in *Nature Climate Change* in June 2018, a group of economists predicted that by 2035 a rapid transition to clean energy could lead to a major crisis for countries whose raison d'être is producing hydrocarbons.[39] The bursting of this "carbon bubble" would have severe consequences for countries such as Canada and Russia, whose economies depend on producing hydrocarbons for a substantial part of their wealth. Quite possibly this crisis would lead to a drop in construction, and therefore to a lower demand for cement and concrete.

That's pure speculation at this point. Certainly those who might look on a financial collapse as the cure to the greenhouse gas problem shouldn't bet on it. What is certain is that China and India, by far the two biggest producers of cement today, experienced little or no impact from either of the breakup of the USSR or the capitalist catastrophe of 2008. However, and the importance of this can't be stressed too much, in the 1990s China changed its path as radically as the former Soviet bloc countries did, but instead of a slump in Chinese cement production, that is the moment when it took off.

In 1990 China's urban population was 300 million, but by 2010 it had more than doubled. Housing all those new city dwellers meant new cities, as we saw in chapter 3, where we talked about home fires. The new cities, along with the infrastructure to serve them, meant an enormous increase in the production of cement, with the concomitant effect on the atmosphere. Between 1991 and 2010, the amount of cement China produced grew by 936 percent.[40] A year later, the country was producing about 1.6 tons of cement per person, four times higher than the historic peak in the United States.[41] And by 2018 China was turning out nearly 60 percent of global cement production.[42]

In India cement production has also soared, but even though it is now the second largest producer of cement in the world, its production is only a bit more than a tenth of China's: 290 million tons[43] compared to China's 2,378 million tons. The uses to which this cement will be put vary also, and that will have repercussions on the atmosphere and more ephemeral manifestations of spirit.

As we saw earlier, an immense part of China's cement goes into making concrete to build its many new cities and to rebuild its older ones. In India nearly 70 percent of concrete goes into the housing sector, but that country has little hope of changing the way people live as completely as China has done. Many factors account for this. Chief among them is the political philosophy on which the two countries are built. China, despite its relatively recent espousal of much more capitalistic economic policies, is still a communist state superimposed on an ancient tradition of central organization and control. When good policies are adopted at the top, they affect the lives of everyone positively. Of course, the converse is also true: when the leadership is wrong the results can be tragic, as occurred in the 1960s when Mao's Great Leap Forward led to perhaps the worst famine in recorded history.

India is a far more unruly union, but since the middle of the twentieth century it has not had a crisis of hunger that remotely resembled those China experienced when its leadership took that wrong turn. Nobel laureate economist Amaryta Sen has argued that India's democratic institutions and reasonably good human rights policies helped avoid this.[44] But the country has missed the boat when it comes to raising the mass of its people out of poverty. Life expectancy and literacy are now both much lower in India than in China because India has not invested enough in health and education to reduce inequality and improve productivity, Sen wrote more recently in *The New York Times*: "For India to match China . . . it needs a better-educated and healthier labor force at all levels of society."[45]

Housing obviously has a role to play in both spheres, although Sen did not mention it specifically. At present there is an estimated shortage of 18 million homes in India, of which 15 million are needed for low-income Indians.[46] In 2015, the government of Narendra Modi announced plans to fill the gap by 2022, and when Modi was re-elected in 2019 he promised to build 29.5 million new housing units in rural areas and 12 million units in urban areas by 2021–2022.[47] But progress is slow. The scheme is based on government grants to people who meet certain criteria, to finance private construction of housing units, including owner-built houses. Determining just who is eligible has proved complicated, though, and the amount of money expended—about c$5.9 billion in the first four years—falls clearly short of the goal.[48] It is also far less than China has put into building new housing: the contrast goes a long way in explaining why China produces and uses so much more cement and concrete.

In the past, India—particularly some Indian states—were more aggressive on the housing front. As mentioned before, Chandigarh, the capital of Punjab, was created out of nothing on the plain below the Himalayas in the years following 1948. As a monument to concrete, it has no parallel aside from Brazil's capital, Brasília: we'll consider both of them in more detail a bit later.

More modest projects have been undertaken elsewhere in India. I saw one of them dating from the 1970s when I visited Kochi in Kerala State on the west coast of India. It is an attempt that does not have many cousins. While researching the book I was working on then (*Green Cities: People Nature and Urban Places*) I visited an apartment in Gandhi Nagar, a development that was one of the first two planned housing "colonies" built by the Greater Cochin Development Authority and which stands on what was low-lying land formerly used for growing rice. The scheme included housing for "economically weak" clients, but also featured a sizeable amount of housing for wealthier folks. My contact's low-rise building sat on a gated lot, with plenty of parking for residents' cars that was guarded by a gatekeeper. It also had elevators, but no built-in air conditioning. Nevertheless, the interior space was arranged so that air circulated through open-work concrete blocks into cool corridors even when the afternoon sun beat down.

But far more common, it seems, is the kind of housing just down the road. It was a completely illegal stretch of small, attached living quarters. They lined the Peradoor canal, a remnant of the lovely waterways that are the signature feature of Kerala State. There I got an impromptu tour from a proud resident who wanted to show me how she lived—and to enlist my support to make the little neighbourhood less threatened by floods during the monsoon season.

The settlement was shaped like an L and lay between a concrete wall that marked the edge of officially recognized development and the canal. Going inside was like entering many other informally constructed developments around the world, ranging from Brazilian favelas to Chinese courtyard houses: intruders are kept out and a community life goes on inside.

One woman invited me in to meet her two daughters and one son. The latter was in x Standard, the last year of secondary school, and seemed delighted to try his English on me. Their house had three rooms: a bed-sitting room that opened off the footpath that served as the street, a middle room with a fan and television, and a third that was part kitchen, part washroom and which looked out on the canal. Inside was dark because the only windows were on the canal side. But outside was bathed in sunlight. Along the wall that formed the front of the houses were many plants, including chilis, marigolds, jasmine, basil, and other herbs. Some were in pots, others in dirt-filled plastic bags. A hose ran along the common wall and women were collecting water. My guide said they had four hours of water and four hours of electricity a day.

The overall impression was of a certain coziness: protection from the outside world, cleanliness in contrast to the trash blowing on the city-approved streets outside, plus sunlight and strong colours: blue of sky and water, green of plants, coloured saris, painted doors, polychrome religious icons.

City officials told me the next day that the little neighbourhood was slated for removal. Not only had it been built without official permission, its location was indeed periodically flooded, sewage was discharged directly into the canal, and the pirated electric connection was unsafe. Today, the settlement appears to have disappeared: it doesn't show up on Google Maps, although the canal does.

The people have moved, obviously. That fact suggests two questions: Where? And would upgrading housing like this bring bigger bang for the buck than the totally new developments that recent Modi initiatives promise?

I have no way of knowing the answer to the first question, and the jury is still out on the second. Relatively low-cost schemes like

Mexico's *piso firme* initiative, which subsidizes putting in concrete floors, have proven that a small outlay can have great positive impacts on people's lives. Regularizing land tenure is particularly important in this kind of intervention because people and governments are unlikely to underwrite plans that spend money where no clear right to build is available. One of the great motors for housing improvement in Brazil, as an example, was a concerted campaign by unions, the Partido dos Trabahaldores (PT), and later the PT governments of Luiz Inácio Lula da Silva and Dilma Rouseff to clarify land title in Brazil's informal housing, including favelas where about 11.25 million people—6 percent of the country's population—live.[49] Lula's government launched *Minha Casa Minha Vida* (My House, My Life), a program to aid low-income, first-time homebuyers, setting minimum standards for housing, arranging affordable mortgages, and clarifying land title. About c$272 billion (US$207 billion) were invested and more than four million housing units were delivered. The program was one of the few that the conservative government of Michel Temer endorsed when it took over in 2017. But when the even more conservative Jair Bolsanaro took office as president in 2019, he effectively axed the program on his first day in office by eliminating the Ministry of Cities, which ran the program.[50]

One criticism of the Brazilian initiative had been that much of the new housing is far away from neighbourhoods where the target beneficiaries previously lived. I heard the same thing in Shanghai at the beginning of China's massive housing makeover in 2006. Ten years later, *The India Times* found 20 percent vacancy rates in schemes that built houses in areas far away from employment or that were incompletely served by infrastructure.[51] Research by three Indian academics similarly found that proximity to a city's central business district increased desirability of housing, as reflected in house prices. Slum dwellers, three-quarters of whom had incomes of less than US$156 per month according to one study, were willing to pay more than US$466 more a year for a house closer to downtown.[52]

Nevertheless, the time may have come for more development of dense neighbourhoods where very small houses take centre stage. At least their promise was internationally recognized in 2018 when

Balkrishna Doshi won the Pritzker Prize, sometimes called architecture's answer to the Nobel Prize. The Indian architect, now in his ninth decade, designed buildings with twentieth-century giants like Le Corbusier, who was a mentor during the planning and construction of Chandigarh, and Louis Kahn, with whom he worked on the gorgeous buildings for Ahmedabad in Gujarat State. His friendship with Kahn lasted for decades—the evening before Kahn died he dined with Doshi and his wife—and Doshi's oeuvre includes more than 100 buildings, including museums and grand government buildings. But it was his work in the realization of Aranya, a modest rural town in a poor area of India, that was cited by the prize-selection jury in particular.

Doshi's plan for Aranya in the central Indian state of Madhya Pradesh was elaborated in 1989. Originally it was to include 6,500 dwellings on a planning area of 85 hectares. Six sectors, each with populations of between 7,000 and 12,000, were to lie to the east and west of a central commercial street. Ten houses, all with a courtyard at the back, would form a cluster that opened onto a street. Septic tanks were to be provided for each group of twenty houses, and electricity and water were to be available throughout. The development would be mixed, in that very basic housing—a concrete slab connected to services—would exist close to housing for more fortunate families. "Housing as shelter is but one aspect of these projects," the jury stated. "The entire planning of the community, the scale, the creation of public, semipublic, and private spaces are a testament to his understanding of how cities work and the importance of the urban design."[53]

Photos of the village, located in a state where a third of the households are considered poor,[54] show brightly painted little houses with banana palms in courtyards and children playing ball in the common areas. Doshi saw that sixty prototype houses were built, but in most cases the owners bought just a serviced lot. Design and material varied, with both brick and concrete block used in the construction. In some cases the houses reflect the rising income levels of the owners: a few three-storey ones now stand on lots where originally a one-room house was built. But as urbanist Rahul Srivastava emphasized after his visit in 2011, the whole plan was to develop the village incrementally.[55] And development certainly has been incremental! It would appear that initial estimates of a 65,000 population for Aranya are far from being met: the 2011 Indian census put its population at 138 families and 785 persons.[56]

Significantly, however, another recent winner of the Pritzker prize, Chilean Alejandro Aravena, also is a proponent of what he calls "half-built houses." The project that the jury particularly lauded was Quinta Monroy, a social housing project of ninety-three houses where most of the US$7,500 budget per dwelling went to buy the land where the residents already lived. What Aravena's group Elemental put up were three-storey structures that included a kitchen, a bathroom, structural walls, and a staircase. The rest of the houses, the half-buildings, were left to the residents themselves to construct, offering double the space normally given to social housing residents.[57] It's notable that the work that Doshi and Aravena have done to provide decent housing for ordinary people was cited by the Pritzker jury. With the exception of the Chinese architect Wang Su, the 2012 winner who designed the Vertical Courtyard Apartments in Hangzhou, Pritzker laureates are usually chosen for their grand public projects or individual houses of great beauty.

Projects like these are far less flashy and much smaller than the high-rise cities being constructed in China, but they may point the way toward another model of providing better housing that is perhaps more respectful of community life, and certainly less concrete-intensive. Population density can be achieved by different models. Take for example the trendy Mile End district of Montreal where the housing type is mostly two- and three-storey attached buildings that are now between 70 and 100 years old. In 2016 it had a density of 12,792.1 inhabitants per square kilometre.[58] That's a lot more than the 2011 population density of Le Corbusier's baby, Chandigarh—9,258 per square kilometre[59]—and even of Beijing in 2010—11,500 per square kilometre.[60] And, interestingly for those who think that Moshe Safdie's idea of "for everyone a garden" is an important principle of urban design, Mile End housing—built on lots that measure 6 or 7.5 metres by 23 or 30 metres—nevertheless usually comes equipped with tiny front and back gardens, or balconies.

Much of that housing was built a century or more ago, at a time when cities were designed to be walkable because there were few alternate forms of transportation for ordinary folk. Housing density made sense then. It still makes sense when governments in emerging nations around the world are trying to provide more and better housing for their people. But even if a lower-tech model is adopted—more small houses on small lots, instead of high-rises—more housing will mean more concrete used. CO_2 emissions will continue to grow.[61]

COLLATERAL DAMAGE, INDUCED DEMAND

All right, that's the direct impact of the production of cement on the air. But it and the concrete made from it are responsible for another major source of CO_2: the gases emitted by the machines used to build concrete buildings and roads, as well as the vehicles that travel on the roads. Transport accounts for between 12 and 20 percent of CO_2 emissions—the figure varies, depending on how the calculations are done and who is doing them—and bringing them under control may well be even harder than reining in emissions from concrete construction.

When my family moved from a small town in Washington State to Southern California, the air most of the time was as clear as it was the day I visited McInnis Cement. I remember how exciting it was to drive south for the first time along U.S. Highway 101 from Los Angeles through Orange County, past thousands and thousands of orange trees with the snow-capped San Gabriel Mountains in the distance. Not far away to the west were coastal communities where fog might roll in most mornings, but by noon it would have dissipated. I've since learned that air pollution was rampant during earlier decades, when orange growers would burn oil in pots around the groves in an attempt to protect the fruit from occasionally freezing temperatures, but that was not an issue most of the time.

There was another threat waiting in the wings, though, or rather under the hood of cars like ours: hydrocarbons that would become smog in the presence of sunlight.[62] The problem had been brewing during the previous ten years, when California's population began to grow massively. But it wasn't until the year we moved, 1951, that experiments began at the recently formed Los Angeles County Air Pollution Control District to see how smog affected people's health: employees of the district actually exposed themselves to smog and recorded how long it took for their eyes to start to tear.[63] The district, by the way, has since been folded into a much larger governmental body, the South Coast Air Pollution Control District, that covers most of the non-desert area of Southern California, north of San Diego County. The lesson to draw from that is that air pollution knows few political boundaries.

Another lesson is that concrete is indirectly responsible for a lot more pollution than that emitted by cement plants. Consider those thousands and thousands of roads built in the last seventy years. They are brilliant examples of the truth behind the adage "If you build it, they will come," and its corollary, "Build it, and the world will be profoundly affected."

In 2005 when I met with environmental officials in São Paolo, Brazil, they were concerned about the effect of new ring roads and diversions that were planned for South America's largest city. The northern part of the Rodoanel Mário Covas had been open for about three years and plans were underway to start construction on the southern segment, which would serve as a corridor for more truck traffic. Up until then, a large part of the 250,000 trucks that travelled each month from São Paulo's industries to Santos, the port on the Atlantic Ocean fifty kilometres away, had to pass through the city's streets or on an inner ring road that was sorely overcrowded. The new beltway segment might cut down on transit time and pollution from idling cars and trucks, the city's urbanists acknowledged, but it would also encourage more urbanization in outlying areas. The result would be more traffic and more pollution within a few years.

That indeed appears to have happened. While a portion of heavy traffic does use the now-completed beltway, congestion in the city is worse, in part because of a great increase in smaller vehicles, including delivery trucks.[64] This added vehicle traffic is what traffic engineers and urban planners call "induced demand." Researchers at the University of California at Berkeley found definite signs of this in an extensive study of traffic on state roads in California using data from 1970 to 1990. In a nutshell, a 1 percent increase in lane-miles resulted in a 0.9

Figure 5.1: Constructed in 2005, this five-level interchange in Dallas, Texas, is an extreme example of how highway engineers have attempted to organize highway traffic in a world where the number of vehicles continues to grow. Photo by austrini, Creative Commons Attribution 2.0 Generic licence.

percent increase in vehicle miles travelled in five years. That is, in the San Francisco Bay Area, Los Angeles, and San Diego regions, each additional mile of a traffic lane on a state highway meant the vehicle miles travelled grew by 12,000 per day (7,440 per added kilometre).[65]

How does this work out in terms of added CO_2 due indirectly to concrete construction? It's estimated that a fuel-efficient car with one passenger would emit 0.1276 kilograms of CO_2 per passenger per kilometre.[66] That means that in the California case, each additional kilometre of a traffic lane adds up to 1309.44 kilograms more CO_2 per day. Multiply that by 365 days a year, then figure out how many new traffic lanes there are in all, and you arrive at a really substantial number.

These are back-of-the-envelope figures, but they are enough to show that concrete contributes far more to the problems of our troubled air than simple calculations about how much CO_2 is emitted by simply making cement: the long-term effects are enormous.

This is not to say that pollution from cement production can't be acute. Using pet coke to produce cement is a way of using a by-product of hydrocarbon processing which otherwise would have to be disposed of, but doing so emits a lot of sulphur and nitrous compounds unless special equipment is in place. In India, pet coke is used, without scrubber or other techniques, to generate electricity and produce steel as well as to make cement, and Delhi, the nation's capital, is choking from the resulting sulphur emissions.[67] As a consequence, in 2017 the Indian Supreme Court directed the national government to either ban the fuel or place restrictions on its sulphur content, and a year later pet coke imports were banned.[68] The cement industry, tellingly, got a pass on the new regulations.

Cleaning pet coke isn't impossible. The fuel contains up to 90 to 95 percent carbon and has a low ash content, which up-to-date equipment can remove.[69] But whether that will happen in India, China, or other countries where it is also used as a fuel in several other industries, is an unanswered question. Chinese researchers who studied its use from 2010 to 2016 noted both the air pollution problems associated with it, and that its use in China had increased by 148 percent. They called for "efficient decontamination systems" and "the clean utilization of petroleum coke."[70]

GREEN CONCRETE

As I said, the air above the McInnis plant was wonderfully clear when I visited. A lot of that had to do with the fact that the plant was still being put into service, but it is supposed to be extremely clean, nearly "green." Spokesperson Maryse Tremblay stressed that the factory will operate to norms that are as much as fifteen times more strict than existing Canadian regulations and that it will produce up to more than 90 percent less sulphur dioxide, 70 percent less nitrous oxide, and 65 percent less dust than other cement plants in Quebec.

What couldn't be seen, of course, was the CO_2, but Tremblay says the company is serious about reducing the carbon footprint by, among other things, using biomass from the surrounding woods as fuel. A year after the plant's start-up, Tremblay told me in an email that studies were underway which could lead to using 100,000 tons of dry wood and forest residue for fuel, thus supplying about 30 percent of the plant's energy requirements and decreasing its CO_2 emissions by 150,000 tons a year. This is predicated on the assumption that wood is carbon neutral, that over time second-growth trees will sequester as much CO_2 as is emitted by the fires that make the cement. It's estimated that the process will take at least twenty years, but as noted before, some forest scientists are skeptical about just how efficient the rebirth process will be.[71]

Whether or not the assumptions are valid, they are used in figuring carbon footprints. Therefore using wood left over from logging operations on the Gaspé Peninsula or grown specifically to be fuel would make it easier for McInnis to meet its obligations under Quebec's cap and trade carbon market scheme.

My visit to the McInnis plant was in late May when, in the cool climate of the Gaspé Peninsula, nature was waking up, taking in carbon dioxide, breathing out oxygen. The fields along Highway 132 to the west of Port Daniel were brilliantly green, and in places the leaves on the poplars and other trees shone as if illuminated from within. Even the mountain of overburden that had been removed to build the plant was coming to life as grasses planted to stabilize the slopes started to grow. A stunningly green landscape . . .

Concrete, in contrast, is not green, usually not in colour, and usually not in impact on the world. The difference was something that haunted me on the trip. How to reconcile despoiling this verdant landscape to make the material essential to the world as we know it; how to somehow mitigate its impacts on the planet, and protect whatever future we have left?

There are attempts going on, I have been glad to learn. Some of them could be qualified as green hype, but others are sincere undertakings which will have—maybe even are having—a positive effect on the air, that basic constituent of the Roman world and of ours.

One North American initiative, the Calera project on the central coast of California, got immense publicity in the early part of the twenty-first century for proposing to turn CO_2 into a chemical cousin of limestone and then into cement by bubbling it through seawater. The recipient of several government grants aimed at providing alternatives to Portland cement, the Calera process "mimics the same chemistry that natural processes use to make strong and tough structures," according to the firm's website.[72] Calera founder Brent Constantz—who calls himself "a serial entrepreneur"—is no longer involved, but is now CEO and cofounder of Blue Planet. This company has had some success in producing aggregate from a process that also changes CO_2 into calcium carbonate, but through a process that is supposedly less energy intensive. It currently sells a line of products including roofing granules, solar-reflective pigments, and purified CO_2 for other uses. "Until recently the only method for offsetting the carbon footprint of your building was achieved by planting trees. With Blue Planet sack concrete the highest CO_2 footprint building material can now be carbon neutral or carbon negative," says its promotional material.[73]

Three other projects designed to attack the problem of CO_2 in cement were among the ten finalists in a four-year competition designed to discover and support start-ups that will turn CO_2 into useful products, the NRG COSIA Carbon XPrize. Funded by firms whose carbon footprint

AIR

is far from neutral—COSIA stands for Canada's Oil Sands Innovation Alliance and NRG is an integrated power company that uses both coal and natural gas for electrical power generation—the C$20 million prize has garnered a lot of publicity, and possibly will point the way to more successful ways to reduce CO_2. Each of the finalists will test their processes by converting CO_2 from either gas-fired plants or coal-fired ones into useful material. CarbonCure, based in Dartmouth, Nova Scotia, aims to retrofit cement plants so they can produce concrete with nano-size mineral carbonate obtained from CO_2 emissions. The CarbonUpcycling UCLA team is producing building materials that absorb CO_2 during the production process to replace concrete.[74] Montreal-based Carbicrete transforms slag from steel manufacturing into precast concrete by curing the material in CO_2 chambers, but pulled out of the competition in 2019 in order to concentrate on actually producing the project in a pilot plant, according to CEO Chris Stern.[75]

Other attempts to cut down on the CO_2 footprint of cement and concrete include a UK project to introduce graphene into the concrete mix, making it much stronger and therefore requiring much less to be used.[76] More conventional additives like silica fume and slag, mentioned before, have already impacted how much CO_2 cement and concrete account for. Then there are the projects which aim to capture the CO_2 during the process of making cement. One promising possibility developed by a research consortium that includes the world's fourth largest cement company, HeidelbergCement, and the Australian technology company Calix is now undergoing tests in Belgium. It has the lovely name of LEILAC (Low Emissions Intensity Lime and Cement) and captures the gas in an almost pure form during the calcining process.[77]

Then there are a number of experiments aiming to replace concrete with wood in large-scale construction. The technical problems previously inherent in building with wood are being solved, architects and engineers in several countries assert. New techniques reduce the risk of fire, and stability issues have been drastically reduced: high-rises made of wood as tall as 18 and 20 stories are in the works in Sweden, while a developer in Vancouver is planning a 35- to 40-storey building.[78] And unlike concrete construction, where coming up with building materials leaves nothing but holes in the ground and CO_2 in the air, wood is a renewable resource, at least in principle. "A forest, if managed properly is a large carbon reservoir," notes a report on tall wood structures.[79] "A typical North American timber-frame home captures about 28 tons of carbon dioxide, the equivalent of seven years of driving a midsize car or about 12,500 liters of gasoline . . . . If Mass Timber building systems were to become common in the building industry, the amount of carbon stored in buildings would significantly increase."

That sounds wonderful: our towns might become the more or less permanent solution to the question of where to put all that CO_2. Indeed, it sounds almost too good to be true, and, unfortunately, it is. Trees planted to replace those cut down for lumber might sop up CO_2 in the air today, but there's a bigger problem: where would the wood to build what people want to build come from?[80]

One of the attractions of concrete in the twentieth century, you'll remember, was that it was a good substitute for wood, which was becoming increasingly scarce for building houses. There's an old Chinese saying that resonates with that: When's the best time to plant a tree? The answer? Twenty years ago. That certainly wasn't done on a large enough scale to provide lumber for a massive shift in construction techniques.

The corollary to that question is: what's the second best time to plant a tree? Usually the answer is given as "now," but that also is simplistic. Not only can tree-planting programs fail because not enough of the seedlings survive, but also with a large enough managed forest, the temptation is to burn the wood for fuel, not to keep it as a "carbon bank," either in the forest or in our houses.[81] No, cutting down CO_2 in the atmosphere is going to require far more action on several fronts.

The gas does have some important industrial uses, among them putting the fizz in beer and soft drinks and prolonging the shelf life of packaged bread and goodies. Most of what's used now is a by-product of making ammonia-based fertilizer, but in the summer of 2018 an unusually large number of fertilizer plants in Europe were shut down for maintenance, causing CO_2 shortages. Brewers and soft drink makers in the United Kingdom had to cut production briefly because they couldn't get enough, a near-disaster in a summer that proved to be much hotter than usual.[82]

Commercial CO_2 capture from cement plants for these ends is a long way off. Nearly all technology advances have been targeted at reducing CO_2 from gas- or coal-fired power plants. Among the most significant ones are a large-scale carbon capture plant in Chinese oilfields that the country announced in 2017,[83] and Petra Nova, the world's largest postcombustion carbon capture system, opened in Texas in December, 2016, and named POWER magazine's Plant of the Year for 2017.[84]

But the ultimate decrease will probably come from Chinese initiatives to control their air pollution problems, because they have such a huge share of the market. Three Chinese companies produced 20.1 percent of the world's cement production in 2018, with China National Building Materials (recently merged with Sinoma) producing 11.6 percent of the total and selling it only in China.[85] Following the Paris Agreement on Climate Change, China announced its Intended National Determined Contributions (INDC)—or how much they're committed to do in the war against climate change. The INDCs include lowering the country's CO_2 emissions per unit of GDP by 60 to 65 percent by 2030 from levels in a baseline year, 2005. To do this it is introducing a national cap and trade program and putting a cap on coal consumption by 2020.[86]

Critics, of course, question China's use of 2005 as a baseline year, since its expansion was well underway by then, and suggest that an earlier baseline would be more appropriate. Others, as mentioned earlier, assert that if China's CO_2 emissions have grown while those of Europe have dropped, it's partly because developed countries have in effect exported much energy-intensive production, with its accompanying CO_2 debt, to developing countries.

China may be preparing to do the same as it embarks on its Belt and Road Initiative. The country, through its government-supported and -controlled industry, is moving toward deep economic involvement in projects elsewhere in the world. An example is the key role Chinese firms and investment is playing in building in the new capital for Egypt. So, while China actually began cutting back on its domestic cement-producing capacity in 2018—the goal supposedly is to eliminate 280 million tons of capacity at home—facilities owned by

Chinese companies elsewhere in Asia and in Africa will be increasing production, according to the industry magazine *World Cement*.[87]

Of course, the quickest way to cut down on CO_2 emissions from cement would be to stop building. How effective that would be was hinted at in the first months of 2020 when concerns about the spread of COVID-19, the new coronavirus, led to the quarantine of several Chinese cities and the shutdown of industry for weeks. As noted before, Chinese business goes on holiday during the Lunar New Year celebrations, but in 2020 the shutdowns were so widespread and prolonged that China's CO_2 emissions in early 2020 were about 25 percent lower than they were during the same period in 2019.[88]

It seems unlikely that slowdown will continue once the COVID-19 pandemic is mastered because the Chinese are determined to increase the country's development. As Charles C. Mann, who has written extensively on carbon capture, says, "The Chinese government faces twin imperatives: lifting people out of poverty and avoiding the worst consequences of industrialization."[89]

Another option for the Chinese would be to switch to the kind of low-rise housing for which Indian architect Doshi and Chilean architect Aravena have been celebrated recently. That is unlikely too: the high-rise "tower in the park" model seems too well entrenched. But building better quality might have long-term positive effects, largely because the housing would last longer and so the future demand for cement and concrete would be reduced.

Observers like Wade Shepard who have studied China's current city-building binge say that housing built in the first wave of construction in the 1980s and 1990s from the get-go wasn't designed to last. The result has been an enormous amount of demolition, with the resulting concrete debris, some of which can be recycled, but a lot of which can't. Whether China's new housing will also be short-lived is an open question. But the same could be asked about buildings going up in North America and Europe.

Agencies like Fannie Mae, the U.S. Federal National Mortgage Association, have looked closely at the expected life of buildings and drawn up detailed check sheets for particular components. For multifamily structures, the only elements that can be expected to last for fifty years are the concrete ones: foundations, concrete slab roofs, and precast concrete panel exterior walls can be expected to last forty-five to fifty years.[90] But the actual wearing-out of a building is not the only reason for demolishing it: "area redevelopment" (34.8 percent) was given more frequently than "physical condition" (30.8 percent) as the reason a building had been torn down in one study of 227 buildings in the United States and Canada.[91]

This brings up myriad questions about the kinds of decisions cities are making—or allowing developers to make—about the way cities develop and evolve. While working on this book, I spent a week in Toronto, which you'll remember has more skyscrapers than any other North American city but New York, many of which date back to the 1960s and 1970s. What was striking was not that these older, mostly residential, buildings were being upgraded: few were that I could see. No, what amazed me was that in the heart of the city some seemingly perfectly good buildings were being demolished to be replaced by more elaborate, taller ones. At that point there were eighty-one skyscrapers higher than 150 metres either under construction or in the planning stages.[92] Housing is expensive and in short supply in Toronto, but a legitimate question is whether this kind of demolition followed by new construction is the best way to fill that need in Toronto or anywhere.

Then there are the sports stadiums, which seem to have a "best before" date of no longer than twenty or so years after construction. Between the late 1990s and 2010, ten were demolished, some of which were imploded while cheering crowds watched.[93] The destruction continued in the new century, but sometimes the old buildings resisted: it took two tries to bring down Detroit's Pontiac Silverdome in 2017. Aside from general decrepitude, the reason usually given for doing this is that the old stadium wasn't equipped for modern technology like super screens or luxury touches.

The stadium built for the 1976 Montreal Olympics is an exception, despite its many structural problems. It was supposed to have a retractable roof suspended from a sloping mast that can be seen for kilometres around, but its many post-tension cables and its fabric roof have been plagued with problems since shortly after it was built.[94] Why it's been allowed to remain is a question I won't go into here, but in part it has to do with two things: Montreal and Quebec politicians like the idea of having a monumental tower that can be compared to the Eiffel Tower in Paris, and to bring it down safely would be horrifically expensive because of the way it's built and its location in a residential area. In the meantime, it draws more than a million visitors a year, and crews are constantly at work maintaining it, including painting the mast, which for most concrete structures is not usually done.

When it comes to skyscrapers, older ones may be more robust than the elegant new super-tall buildings. Take, for example, the difference between the Empire State Building and the Burj Khalifa in Dubai, currently the world's tallest building. Fourteen people died when a B-25 airplane crashed into the Empire State Building in 1945, but the building, completed in 1931, reopened for business a few days later. "Back in the early 20th Century they were still calculating everything by hand, so they always added extra steel just in case," structural engineer Roma Agrawal told the BBC. Because of this, he said, the Empire State Building, which is less than half the height of the Burj Khalifa, weighs two-thirds as much.[95]

That's over-engineering, but there are also flagrant examples of under-engineering that sometimes come with tragic results. One occurred in Bangladesh in 2013 when the Rana Plaza, an eight-storey commercial building, collapsed, killing 1,134 people. Most of the victims were garment workers who couldn't get out of the sweatshops in time when cracks in the structure, noticed the day before, suddenly began to widen. Subsequent investigation showed that poor quality materials, a foundation laid on a filled-in pond, and the addition of three floors more than had been approved contributed to the collapse.[96]

But, just as McGill's Saeed Mirza says that the rapid deterioration of the Champlain Bridge shouldn't condemn concrete as a material to be used for big public works projects, the Rana Plaza disaster shouldn't be an argument against concrete being the ubiquitous building material of our time. What is at issue here is the practice Mirza decried of "design it, build it, forget it," its first corollary, "don't ever check to see if the rules are being followed," and their opposite, "maintain the damn thing."

Maintenance is the key to making modern concrete the material that is truly the "Rock of Ages" it was thought by many to be in the early and mid-twentieth century. Maintenance is no more sexy than cleaning bathrooms, but like housework it can make the difference between a livable place and a slum.

As art historian Adrian Forty writes, "The early pioneers of concrete thought they had found an everlasting material; in this they were to be sadly disappointed, for . . . it is not as stable a substance as they had supposed. Although it may last for a very long time, it does undergo changes and according to the precise combinations of ingredients and local atmospheric conditions, may lose its strength or otherwise deteriorate over time. Among concrete experts, the view is that all concrete structures will sooner or later need radical repair—a difficult and costly process."[97]

Maintenance also seems to have helped some of the Roman concrete structures like the Pantheon weather centuries rather well. The excellent concrete used to form its dome is partly responsible, but also important was what happened after the fall of Rome. The building became a Christian church in 609 CE, which meant that it was not abandoned for any length of time, not plundered, and looked after rather well over the centuries. In addition, the original wisdom of its architects can be seen in the way that fissures that have developed over time seem to have had little effect on the soundness of the building.[98] It has aged, but the damage is nothing compared to what many more recent concrete structures have experienced, even though rain and snow can enter from the opening at the top of the Pantheon's dome, the oculus.

In contrast, "time and weather, which give mellowness to brick and stone, make untreated concrete more and more dirty, dark, and untidy and rapidly lower its initially low power of reflecting light," warned the Royal Institute of British Architects after World War II when plans were being drawn up to reconstruct the war-damaged country.[99] Yet ugly stains may be nothing more than blemishes, with real damage to reinforced concrete not being immediately apparent. That occurs when water enters and attacks the steel inside, silently weakening the structure.

To address problems with modern concrete, the Getty Conservation Institute has undertaken a huge project aimed at conserving some of the twentieth century's most remarkable concrete structures.

In 2017 three museums designed by Le Corbusier, two in India and one in Japan, received grants to explore ways to ensure that the buildings remain structurally sound and as beautiful as Le Corbusier would have liked.[100] As the project's mission statement says:

> Reinforced concrete was the material of choice for many architects of the modern era, and they exploited the material in a multitude of creative and innovative ways . . . . The current state of decay of many significant reinforced concrete structures arises from the novelty of the material and construction techniques used at the time of their construction, and the fact that architects and engineers experimenting with reinforced concrete often pushed the limits of the material structurally and architecturally.
>
> Although there are many well-constructed, carefully crafted concrete buildings of this time, there are also many buildings suffering rapid deterioration due to poor quality materials or construction. This is often the result of building at a time when materials were scarce, under pressure for accelerated construction, and with little quality control . . . . Moreover, these buildings often suffer from the mistaken belief that reinforced concrete was a maintenance free, extremely durable material. The result is a large stock of culturally significant reinforced concrete buildings with deterioration that ranges in scale from local to general.[101]

However, it is possible that the human spirit, that other component of air, can provide solutions for these problems. Certainly, it has long used concrete as a medium for expressing wonderful and extravagant thoughts.

CONCRETE BEAUTY, BEAUTIFUL CONCRETE

When I started thinking about concrete, I admit I had a prejudice against it on aesthetic grounds. The material was useful, essential even, but I could not immediately think of any example of a concrete structure that was beautiful. The scales began to fall from my eyes— or, to use a concrete metaphor, I began to see the truth as the form-work was removed from the beauty that had been created—after a talk with a young friend. A building contractor who specialized in custom residential jobs, he was enthusiastic about the wonders possible with concrete. Any shape, any surface, any thickness, any height: the builder's creativity knew no bounds, he said.

And he was exaggerating only a little. There are technical limits to what concrete can do, as we've seen, and even within those bounds, skill and knowledge of the material are necessary to achieve the desired effects. But what is possible is astounding.

It should not be surprising that some of the most striking architecture created is done for places of worship. Evidence is piling up that the desire to meet and commune with higher powers may be the major reason why what we now call cities were formed. For decades archaeologists and social historians speculated that agriculture came first in human history, that our hunter-gather ancestors settled down to farm and then, because they had more food, could allow themselves the luxury of making special structures for collective activities. But a 12,000-year-old settlement in what is now Turkey has upended that idea. The local population call it Göbekli Tepe, or Potbelly Hill, because of the way it rises above the plain: obviously they had no hint that underneath lay a temple that is older than any other discovered so far anywhere in the world. Excavations have turned up no evidence that the place was a settlement, an early city, a place where agriculture or pottery making were practised: there are no remains of housing, only items like statuary, artwork, and things that could only be used for some sort of rituals. These include stone vessels big enough to hold more than 150 litres of liquid, and mountains of broken animal bones that are perhaps the remains of feasts fraught with symbolic meaning.[102]

The find's implications are that monuments and sacred places lie at the core of our social organization and date back to before the first

Figure 5.2: Reinforced concrete takes up where the architects of Gothic cathedrals left off in the Église Notre-Dame du Raincy near Paris. The material was chosen for cost reasons, but it allowed for slender supports and huge stained-glass windows. Photo: Paul M.R. Maeyaert. Licensed under Creative Commons Attribution-Share Alike 3.0 Unported licence.

of us settled down to farm. It should not be surprising, therefore, that many of our most beautiful constructions have been intimately connected with belief systems, from Stonehenge to Gothic cathedrals.

With the signal exception of those built by Romans, these buildings were almost exclusively made of stone until the late nineteenth century. The example of what the Romans had accomplished in such magnificent buildings as the Pantheon was not pursued when modern concrete came of age. Part of this was snobbishness, as we noted before: concrete was not a "noble" material like stone and so was not thought suitable for buildings of important religious or symbolic meaning. The delay in utilization of concrete for big projects also was linked to the architectural thinking dominant at the end of the nineteenth century—and it should be noted that in modern times architects, unlike the unnamed geniuses who built the pyramids and Chartres for example, have frequently tried to explain what they were doing. Influential architects in France particularly espoused "structural rationalism," the idea that a change in materials required a change in form.[103] Reinforced concrete was new, and therefore it really couldn't be used until some new way was found to use it, it was argued.

Nevertheless, by the early twentieth century some architects saw that the material could be exploited for exploring the aesthetic of Gothic cathedrals. The steel bars reinforcing concrete made it stronger than the stonework of the Middle Ages and so concrete could "fulfill the dreams of the Gothic builders . . . who were not able to erect as many nor as high towers on their cathedrals as they planned."[104] Chief among these new creations was Notre Dame du Raincy, a Parisian church completed in 1924, where the strength of the material allows slender interior columns to support a vaulted ceiling, leaving the walls free for a splendour of stained-glass windows.

Concrete has been used increasingly ever since for religious buildings of many kinds. I saw brilliant examples of them on my visit to the Brazilian capital, Brasília. The first was Oscar Neimeyer's iconic cathedral, properly known as Catedral Metropolitana de Nossa Senhora Aparecida, on the Monumental Axis that forms the backbone of the carefully planned city. Set on a square paved with huge concrete blocks, it is circular and topped by a framework of white ribs curving first inward, then outward almost as sinuously as the waist of a woman. Stylized statues of four men stand in front of the structure, as does a bell tower that looks like giant candelabra. The statues are of Matthew, Mark, Luke, and John, a visitor will learn, but aside from a cross that springs upward from the middle of the roof, this does not look much like other Christian churches. The entrance is not through grand doors, but underground through a tunnel. As you walk inside, your eyes have just enough time to become accustomed to the low light when—wham!—you find yourself in the glass-roofed sanctuary. The light flooding in dazzles, so the statues of angels hanging from the ceiling are at first lost against the firmament.

The second concrete church is not far away in one of the superquadras, the neighbourhood-sized apartment-block complexes with a population of 8,000 or 10,000 souls. Also designed by Niemeyer, the

Figure 5.3: Brazilian architect Oscar Niemeyer designed the Catedral Metropolitana de Nossa Senhora Aparecida for the Brazilian capital, Brasília. It is an example of concrete at its most beautiful, in a city designed around the automobile. Photo: Mary Soderstrom.

Figure 5.4: Inside Brasília's cathedral: a triumph of light in concrete. Photo: Mary Soderstrom.

Igrejinha de Nossa Senhora de Fátima, the Little Church of Our Lady of Fátima, is partially open to the outdoors, sheltered by a triangular roof of concrete that curves upward like the coif of nun. The noontime I was there it was empty of worshippers, but it was as surprising in its originality as is the cathedral.

The Japanese architect Tadao Ando also uses light and concrete in many of his spiritual creations. His Christian Church of the Light, built in 1989 in Osaka, takes its name from the slits cut in one wall of the concrete structure that forms a cross anchoring the small sanctuary. More recently, his setting for a giant Buddha in Sapporo makes a nod, perhaps, to Niemeyer's cathedral by surrounding the statue in earthwork where access to the sculpture itself is through a vaulted concrete tunnel. At the end one sees the Buddha silhouetted against the open sky, where natural light looks for all the world like a halo.[105]

American architect Louis I. Kahn also played with light when he created the mosque that is the entrance to the legislative complex in Dacca (now Dhaka), Bangladesh. In 1970, shortly after the building was finished, he told *The New York Times*, "I observed that the Pakistanis pray five times each day, and earnestly. And the Assembly Hall is a transcendent place where no matter what kind of rogue you are, when you go into the Assembly, somehow you should vote for the right thing. In the program there was some semblance of a prayer hall, which was 3,000 square feet, with a closet for prayer rugs. But I turned it into 30,000 square-feet—and prayer rugs out all the time." The result is a chamber where light streams indirectly down, in hopes that it will illuminate politics as well.

The problems and triumphs of working with concrete are brilliantly demonstrated in Kahn's work on this complex, begun in the 1960s when mostly Muslim Pakistan was divided into eastern and western portions on either side of the great mass of mostly Hindu India. It was finished about ten years later, following a civil war after which East Pakistan became Bangladesh.

Like Ando's work, Kahn's work showcases the importance of two ingredients that mean the difference between a mediocre concrete building and one that borders on the sublime: the formwork that constrains the concrete after it is poured, and the aggregate mixed with the cementitious material and water.

The forms my father used back when I was a child were cut from sheets of plywood and held in place with whatever spikes he could find lying around or borrow from his friends who were also upgrading their unpretentious houses. But that kind of slapdash affair has never had a place in professional construction: historian Adrian Forty emphasizes that one of the big skill shifts in the industry came when stone masons were replaced by carpenters. Instead of knowing how to shape stone, what was necessary for the men who did the building was to understand how to build forms strong enough to hold concrete yet that could be could be stripped away after twenty-four to forty-eight hours to be used again. They also had to know how much reinforcing steel to put in place before the concrete was poured, and we've seen how at first that was more by guess and gosh than by calculation.

Kahn's Pakistani project ran into problems finding wood for the forms since plywood was unavailable. He and his team had to look to a furniture factory about 100 kilometres away where a mahogany-type wood could be processed and cut. Then came the problem of pouring the concrete into the forms. When the work started, building with concrete on the Indian subcontinent didn't mean rolling up to a construction site with ready-mix in a pumper truck. Rather, workers carried the concrete on their heads, one panful at a time, from a cement mixer, across rough terrain, and then up bamboo scaffolding. In a day, they could place a wall about 1.5 metres high, and almost inevitably, when they began their work the next day, there would be a line separating the two pours. Kahn and his crew were not pleased by this, but there seemed no way around it until Kahn hit upon a brilliant way of incorporating the "mistake" into the finished building by placing a thin strip of marble at each line of demarcation. The result is an elegant play of light against the darker concrete. (The colour, by the way, was imposed on Kahn because only a regulation Soviet concrete was available in large enough quantities at that time in Pakistan.)[106]

For the Salk Institute for Biological Research near San Diego—considered by some to be Kahn's most successful project—there was plenty of plywood for the forms, but the usual way of using it wouldn't work. Instead of simply oiling the forms to make them easier to remove, in order to get the smooth finish desired the forms had to be coated with a special resin, which presented its own problems. It also took several tries before the right reddish sand was found to produce the desired tint to the concrete.[107]

Ando's work, like that of many Japanese architects, is noted for its smooth textures and precise formwork. The techniques aren't secrets; the moulds are carefully prepared and then they are hammered repeatedly with wooden mallets during the pour to stimulate the flow of concrete and release air bubbles: obviously air entrainment to increase resistance to freeze-thaw cycles is not considered important. After the formwork is removed, the surfaces are often polished and treated with a glaze, notes Forty.[108] Even when the finished concrete shows the imprint of the formwork, the effect is well thought out, often hinting at the character of the wood that made the forms. Forty says that this is a reflection of the high place that wood has enjoyed in traditional Japanese architecture. Concrete has advantages over wood in that it is fireproof and more resistant to earthquakes than traditional Japanese construction techniques, but wood remains revered.

When it comes to big new buildings today, though, wood forms are pretty much a thing of the past, even in Japan. Various systems of metal formwork have been developed which can be hiked up one floor at a time as the building rises. An example is the 306-metre, 80-storey Cayan Tower in Dubai, which spirals over 90 degrees as it rises. Aluminum formwork used in the structure—which can be reused up to 500 times—was rotated 1.2 degrees clockwise from the floor below each time it was hoisted upward to frame a new pour.[109] The result is a slim, shining tower that appears to turn slightly, almost as if it were dancing. It stands out from the more conventional skyscrapers that crowd Dubai, rising higher and higher as the oil-rich country's fortunes soared. It is not unique, however: giant buildings like the Evolution Tower in Moscow, The Turning Torso in Malmö, Sweden, and The Shanghai Tower in Shanghai all are built to spiral as they rise. They are examples of a sort of architecture composed of some truly monumental buildings that belong to a school of architectural thought loosely called "contemporary."

A word here about the idea of "schools" of architecture: sometimes architects are grouped, or group themselves, into schools. Each school usually has a body of theory attached to it that may or may not reflect what is actually going on when buildings are designed. Understanding what architects mean when they try to explain what they're doing can be difficult. It's a truism that many of them are far more gifted spatially and artistically than they are verbally. An example is one of Louis Kahn's famous quotes, which is hard to puzzle out: "You say to a brick, 'What do you want, brick?' And brick says to you 'I like an arch.' And you say to brick, 'Look, I want one too but arches are expensive and I can use a concrete lintel.' And then you say: 'What do you think of that, brick?' Brick says: 'I like an arch.'"[110]

Which means what? That it's good to know what a material is best suited for? Perhaps.

Brush away some of the hyperbole and inexact phrasing and what you come up with is the fact that during the twentieth century two architectural tendencies were discerned by observers. The first was called the International Style. In theory it eschewed ornament and colour, used lightweight, industrial materials, and revelled in modular forms, flat surfaces, and glass, with volume emphasized instead of mass, according to the Getty Research online thesaurus.[111] Closely related to it is the modernist style—in fact it's sometimes hard to tell which style a particular building or architect represents.

The second major style is Brutalism, which, despite the menace of its name, refers to the French term for raw concrete, *béton brut*, and not to any intentional oppressiveness. Indeed, many of the forces behind this sort of architecture are idealistic: the desire to make good and liveable buildings for people to work, live, and play in. But here again the lines are blurred: some say that Le Corbusier's Unité d'habitation in Marseilles is Brutalist, while others would classify much of his work as in the International Style. To complicate the matter, architects working in the International/modernist style also frequently used reinforced concrete for their buildings' basic structure. A crucial difference may be that in International/modernist buildings, concrete was more often used to enclose space—the curving walls of Niemeyer's cathedral in Brasília is a case in point—

while in Brutalist buildings the mass and texture of concrete was played up.

Montreal's Habitat 67 is sometimes called Brutalist, by the way, but architect Moshe Safdie objects because the concrete surfaces, while massive, are very carefully finished. "I wanted Habitat to look like a highly finished sophisticated project, which is why I went to pre-casting and I worked very hard to get formwork that would give you really a smooth machine-like surface rather than a brutal rough surface and the boxes come together as very pure geometry," Safdie told Anna Winston of *Dezeen* magazine shortly after he won the Gold Medal of the American Institute of Architecture in 2015. "So I think of it as an anti-Brutalist building, a reaction to Brutalism—it just happened to be built in that period, but it wasn't a Brutalist building."[112]

Today, perhaps it is simpler to define anything that is being built currently as being just "contemporary." This descriptive term also plays down the idea that the geniuses of architecture set the tone for what we build. Rather than putting architecture in semantic boxes, it is better, I think, to talk about four sorts of builders as being responsible for our built environment. The first are planners who lay out the streets and see that the water, power, sewage, and other services are constructed. Sometimes invisible, they're the ones who set the framework for how cities, towns, and even nations develop.

The second are what you might call the "normal" constructors: the contractors, business people, and developers who work within the urban framework to build housing, factories, office buildings, hospitals, you name it. You might know the names of those operating in your neighbourhood because they could be the folks you elected, or because the local real estate tycoon plasters his or her name all over, or because the contractor whose pickup is parked across the street has his name on it.

The third are ordinary folk who live and work in the buildings. In many places, like the acres and acres of informal housing in India or South America, they are the people doing the building. In more affluent areas, they are the "market," the ones who vote for architectural ideas by choosing to buy or rent dwellings that embody them.

The fourth sort are the architects who may be involved in only a small number of projects, but whose ideas can ripple out through society until we are all affected. Le Corbusier was one, as we saw in chapter 3, "Fire" (although it should be remembered that many of his ideas had already been tried out in Sweden and the USSR). It would be very hard to imagine the enormous housing construction boom in China today were it not for Le Corbusier's "tower in the park."

Louis Kahn's influence is wide ranging too. One of his most successful projects is the Kimbell Art Museum in Fort Worth, Texas, which features five low, connected structures with barrel arches as roofs. They look a bit like wartime Quonset huts, but according to Wendy Lesser, Kahn's biographer, that model was not his inspiration. Rather it came from cycloid arches made by rolling a pencil attached to a wheel down the edge of a piece of paper, a moving circle.

A good idea doesn't stop rolling along, though. In 2012 Spanish architects won a competition to build an art gallery in Kabul, Afghanistan, that looks very much like the Kimbell, a resemblance that did not pass unnoted on architecture websites.[113] Then, in 2018, Zaha Hadid Architects won a contract to build a school in rural China that looks amazingly like both the art museum projects. The school is projected to serve 120 students in a rural area of Jiangxi province. They'll live, study, and play in a dozen parabolic vaults constructed from concrete.[114] But unlike the two museums, the vaults will be made from concrete placed in forms made by a robot which will cut foam board into the proper shape. This illustrates several things: that new techniques are liberating concrete from earlier restraints and making complicated designs possible in places where they wouldn't have been; that China, with its immense building program, is in the forefront of innovation in concrete; and that ideas travel and can be adapted to many situations.

Nowhere is that clearer than with the work of Moshe Safdie.

More than fifty years after Habitat 67, Safdie is enjoying a moment of recognition that has been a long time coming. He has said that the fact he was unable to get similar projects off the ground was a big disappointment. "Habitat was only about a fifth of the project I proposed, and if the whole thing had been built, I think that 20th century architecture would have been profoundly different," he said in an interview with *Architecture Daily*.[115] As it was, he spent a career designing some gorgeous public buildings in the United

States, Canada, and Israel. Not all of those projects were built—he says only about 50 percent, adding that it's important for an architect to build, not to design just for the sake of designing. When the time came and the opportunities arose in China, Singapore, and Sri Lanka, he was ready to return to many of his core ideas of density, mixed uses, and access to greenery and gardens. His Sky Habitat in Singapore, for example, features thirty-eight floors of apartments, all of which have views and terrace or balcony gardens, in two towers connected by sky "streets," much like the walkways in the original Habitat. But this is far from being housing for the masses: in 2019 apartments there were selling for about twice the price of standard Singapore HDB flats. They're not luxury housing, though, Safdie told me in an email, because of the project's location on the edge of Singapore, which is not where the wealthy live.[116] (His Marina Bay Sands complex, on the other hand, definitely is top of the line. Its distinctive silhouette—three towers filled with casino, hotel, restaurants, and first-class restaurants, with a SkyPark on the top—formed the backdrop for much action in the movie *Crazy Rich Asians*, symbolizing the city-state's wealth and glamour.)

Two of the elements that Safdie hoped to create in the original Habitat 67 have become commonplace, he said in the *Architecture Daily* interview: high density and mixed use. The sizes of his projects in Asia are big enough to profit from economies of scale and are constructed on large enough sites so the buildings can be placed with great care. This has resulted in "a paradigm shift for our practice," he said. After Habitat 67, most of his work in the United States was focused on institutions because "reproducing, replicating Habitat was not a simple matter. There was so much resistance to building technology, building prefabrication."[117]

Shortly after he'd won the American Institute of Architecture's Gold Medal, Safdie talked to *The New York Times* about his career. He was enthusiastic about the linear park made of the old elevated railroad, the High Line, in Manhattan, but he said that it was a shame that luxury high-rises are being constructed nearby without much concern for each other's views or ambient light. "Every developer wants to have his tower smack against the High Line," he told the *Times*. "The quality is sure to be higher than average, but if you had some kind of thoughtful urban planning, it would be greater still."[118]

When he spoke, his firm had just announced plans for a Habitat-influenced development, a sixty-four-storey mixed-use building in Midtown Manhattan with public gardens, balconies, and careful consideration of light. At the time, he told the *Times*, he was "still finding ways to adapt, modify and expand on his original theme: 'For everyone a garden.'" That notion, he added, is "a metaphor for making an apartment in a high-rise structure into what connotes a house."[119]

Perfectly consistent, of course. But the New York project slipped away from him and his associates: the developers hired another architect in the fall of 2017 to design a building that is taller and more conventional than what Safdie was proposing.[120]

The setback illustrates how difficult it can be to transform an idea, a vision, into reality, how hard it is to make concrete something as seemingly unsubstantial as air.

For another example of that difficulty, go back in time more than thirty years before my visit to McInnis Cement—to Paris on May 25, 1983.[121] Those assembled are waiting for the announcement of the winner of an architectural competition to build a monumental building at the end of Paris's Axe historique, which runs from the Louvre, that ancient fortified castle that now is one of the world's most visited museums, to La Défense on the other side of the Seine to the west. Inside the new building, whatever design is chosen, will be space for a centre for innovation and technology. But the main criteria for choosing the winner are beauty and readiness by the summer of 1989, when France would celebrate the 200th anniversary of the French Revolution.

Nearly 900 architects had entered the open, blind competition, sending in sketches which were not identified except by number. After an initial winnowing, a jury of what today might be called "starchitects" invited five of the teams to produce models and detailed plans.

This afternoon at the Elysée Palace, in the presence of the French president François Mitterrand, a handful of highly placed government

Figure 5.5: La Grande Arche outside Paris is an example of a vision made concrete. Photo: mshch/ iStock Photo.

officials, and architectural giants, one of them opens the envelope with the slip of paper identifying the name and number of the winner. It is Number 640, entered by Johan von Spreckelsen, architect, assisted by Erik Reitzel, engineer, both Danish.

Their entry is an immense, hollow, white cube to be made of concrete covered with marble. The dimensions are such that the Notre-Dame de Paris cathedral could fit inside the arch formed by the sides and roof of the cube. It will be clearly visible from the centre of Paris. It will dominate the long perspective that passes through the Arc du Caroussel to the Arc de Triomphe to La Défense; no one will be surprised when it eventually is christened La Grande Arche.

Von Spreckelsen's cube is beautiful and pure. It is exactly what is needed, the press will say. It will transform and bring order to La Défense, which has become Paris's business centre and is now crowded with skyscrapers that have been forbidden to be built in the inner part of the city.

But no one has ever heard of the architect or engineer. Even a quick phone call to the Danish embassy doesn't give more information because both von Spreckelsen and Reitzel are unknown there. And when French officials call the telephone numbers on the entry, they find that von Spreckelsen has gone fishing and can't be reached, while Reitzel initially doesn't want to talk because his family is celebrating his son's tenth birthday.

It is a fairy tale where unrecognized genius is suddenly appreciated, where merit is rewarded, and beauty wins out over the ordinary. There are certain similarities with Moshe Safdie's story and Habitat 67: a brilliant design comes from unknowns and the world is wowed.

But conceiving a design is one thing; building it is another. As one of the people involved in seeing the idea transformed into reality commented, in von Spreckelsen's "spirit, everything should remain light as air . . . . But making something light from his idea isn't possible. So much depends on the dimensions, on the size of the support beams . . . the thickness of the concrete floors, the kilometres of ventilation, all that sort of thing."[122]

"He was a poet, the Arch is the work of a poet," says another person who worked on the project. The implication is clear: unless poets can work with others who are more competent when it comes to practical matters, a project risks being a failure.[123]

So technical details are passed on to others. Von Spreckelsen concentrates on the artistic ones, like where the marble cladding will come from in Italy. He does not get involved in the technicalities of how to make the concrete super beams to support the roof: the challenge is immense, since they will be 110 metres long, 9 metres high, weigh 2,000 tons, and must be poured in place, 100 metres in the air.

Then the government changes, and the budget is reined in. Some major modifications have to be made for budgetary reasons, and also because of technical problems.

Von Spreckelsen resigns because, it seems, he can't bear to see his vision of the building modified. Two years before La Grande Arche is inaugurated with fireworks and pomp, he dies. Some would have you believe that his heart was broken, literally as well as figuratively, that he was in effect sacrificed for the success of the monument, as surely as if he had been buried in the foundation of his building, engulfed in concrete. At least that is what was suggested by Laurence Cossé, who tried to write a history of him and his grand project.[124]

It is a tragedy, a cautionary tale of the dangers of attempting to soar too high, particularly, perhaps, when your vehicle is as solid as concrete.

But, as I said, ideas sometimes circulate from place to place, as does the air. Like seeds, they may be buffeted by winds for a long time before they are blown to fertile ground. Look at what Moshe Safdie is building in Asia now. After decades when his dream of "for everyone a garden" appeared unrealizable, it now is taking form. His linked towers with their individual gardens, their arching passageways thirty storeys up, and their open spaces letting in light, continue his ideas of fifty years ago.

Then there is Golden Dream Bay in Qinhuangdao. Safdie says that, unlike the several architects designing buildings along Manhattan's High Line that compete for light and views, he and his team wanted to create "urban windows" that wouldn't block the vista. Their solution to the problem was to build vertical modules of 20 storeys that on one side are stepped back to allow rooftop terraces for individual apartments. Two buildings are spanned by slabs from core to core, making gigantic "windows" in each cluster of buildings: one opening is 31 metres wide by 47 metres tall, and another is 35 metres by 40 metres.[125] Each has the monumentality of La Grande Arche, translated into another context where the human scale is found in the details and, thank goodness, the architect lives to see the realization of his project.

Sound far-fetched? Perhaps, but consider this: among the details that von Spreckelsen wanted in his perfect concrete monument was a representation of *nuages*—clouds—in the centre of the arch. His original idea was to use glass, but because of technical problems, the clouds ended up being tent-like billows of fabric suspended from the inside of the arch. Not very satisfactory, most agree. When I visited the structure I, like many others, thought the installation was a temporary shelter of some sort, maybe tarps put up to keep spectators dry during special events in rainy weather or to protect passersby when

Figure 5.6: Night-time at Golden Dream Bay in Qinhuangdao, China, where clouds are projected onto the underside of the walkway to look like the sky. Photo: Tim Franco. Used with permission of Safdie Architects.

workers were doing repairs. But that wasn't it, it was just an idea that hadn't found the proper medium for expression.

And then I came upon a promotional photograph of one of Golden Dream Bay's "urban windows." Taken in early evening with the deepening blue sky in the background, it shows the overhead passageway connecting two buildings with a mottled pattern of blue and white projected on its underside. It looks like the sky, like the clouds.

Safdie says there is no connection with von Spreckelsen's ideas, that the light projection is simply "a kind of decorative feature proposed by the lighting consultants."[126] But it also might be seen as the unconscious embodiment of that most ephemeral of things, an idea— an idea whose time has come, carried to the other side of the world the way air moves around the globe, and then made concrete in all senses of the word.

CHAPTER 6
THE END OF THE ROAD

When I went to the World of Concrete trade show I wasn't quite sure what I would encounter. It was a year, almost to the day, since Donald Trump had been inaugurated as the forty-fifth president of the United States, and political discourse in the United States was becoming more and more fiery.

In November 2016, Nevada voters had chosen the Democratic candidate, Hillary Clinton, by 47.92 percent, versus 45.5 percent for Trump. Nevertheless, I expected to pick up a lot of pro-Trump vibes at World of Concrete, since many of the people involved in construction are good old boys, the solid base of Trump support. My expectations were only reinforced when those of us who took the field trip to Hoover Dam were given red visors that looked mightily like Trump's "Make America Great Again" baseball caps.

But, surprisingly to me at least, I only saw one overtly political statement while I was there. It was part of the display by the maker of huge-but-portable concrete batch plants, and exhorted everyone to ask their representatives, senators, and the president what had happened to the massive infusion of funds in infrastructure programs that Trump had promised during his campaign, and which, it should be noted, had not been approved by early 2020.[1]

Of course, if I'd gone drinking with some of the guys I might have got an earful, but old ladies like me don't usually do that. More to the point, people in the concrete business know that a great amount of their work depends on political decisions. Even in times and places where it seems development and construction operates in a "free market," this influence is enormous. Take the North American construction boom of the 1950s: without government-guaranteed cheap mortgages and government-financed infrastructure programs, the millions and millions of suburban houses would never have been built. Similarly, the housing bubble that ended in the Great Recession in 2008 was built on lending policies that governments tacitly approved, even if they shouldn't have. There is no doubt that behind-the-scenes lobbying and political donations are probably more effective than public confrontation, whether in a democracy where the decision makers can change every few years or in a one-party state.

This has been true since well before Roman emperors decreed big concrete projects. That was the case in the days of the New Deal dams, it was true in China, where the world has been rebuilt in concrete, and it is true now elsewhere in the world as China spreads its influence around the world with its Belt and Road Initiative. Undoubtedly it will also be true in the future as governments around the world try to deal with the ravages brought on by what now seems inevitable climate change. Making wise decisions is the biggest challenge that we collectively have ever faced.

Shorter term decisions will affect what happens at Port Daniel-Gascons. Maryse Tremblay and other spokespeople for McInnis have insisted all along that the plant will be profitable, pointing to projections of unmet demands for cement in the United States and to the weak Canadian dollar, which could give McInnis Cement a comparative advantage. But that favourable position could change if the United States puts tariffs on cement, or if Covid-19 leads to a major economic depression. Then there is the matter of the immense cost overruns incurred during plant construction. For a while it seemed that the Caisse de dépôt et placement du Québec, Quebec's pension fund that now controls McInnis, might be thinking about bailing out. Just about the time I went to World of Concrete the usually reliable Bloomberg news service reported that the Caisse was looking for a buyer for McInnis Cement.[2] The deal did not materialize, and in 2019 Maryse Tremblay said that everything is copacetic after a refinancing deal arranged by the Caisse in July 2019 that injected c$500 million in private capital into the project.[3]

All that seemed a long way off the morning after my visit to the plant, however. My husband and I decided to continue on the highway that runs past the installation, Quebec 132. It circles the Gaspé cap Peninsula, running past fishing villages that have been there for a couple of hundred years, beside farms that produce root crops and pasturage but not much more because of the short growing season. On its southern stretch, the highway skirts the forest that covers much of the peninsula's high ground, and which has been extensively logged. What is growing back in the hinterland may do something to redress the CO_2 balance in general, but there are no calculations made so far that might assure that.

After leaving the town of Gaspé at the end of the peninsula, the road cuts across high ground to come down on the peninsula's north side. This is the beginning of the main part of the estuary of the St. Lawrence. The river is still fifty or sixty kilometres wide here, and the mountains on the opposite side are just barely visible.

Then at the bottom of the grade the first sign appeared, signalling a hazard that comes from concrete: we were in for seventy kilometres of rocks on the road and difficult conditions. The sign didn't say why, but it soon became clear that the road was being squeezed between a high cliff and the waters of the gulf.

Building the road there was probably an error from the beginning, but climate change is making conditions worse. It was nearly high tide on an ordinary day when we drove along, but nevertheless waves were lapping almost at the edge of the breakwater. Storm winds and high tides had shut the route down for days the December before we made the trip. Vehicles were required to make sorties onto back roads on the highland south of the river that amounted to detours of more than

150 kilometres. Something similar had happened six years before, and scientists are sure that more damage awaits.[4] It was easy to imagine what will happen as sea levels rise due to climate change. The broken blocks of concrete and fallen boulders that had been dumped in particularly vulnerable places along the shore were not going to be much protection. The irony of using concrete to save a road threatened by forces unleashed in large part because of concrete was apparent to me, but perhaps not to Quebec's highway authorities.

Out to sea we saw a few ships passing in the ship channel. No cement bulk carriers yet at that point, but shipments of cement from the McInnis plant would enter the river system here when production was in full swing.

Jacques Cartier didn't get this far on his first trip, the one when he put into the Baie de Chaleurs and stopped at Port Daniel in June of 1534. He returned to France at the end of that summer, and then set out again the following May. This time he found the main channel of the St. Lawrence, which he hoped was a passage to the riches of the Far East, but he was stopped at the rapids where Montreal stands now. His ships could go no farther up the river. Today, you can see this first set of rapids just outside Moshe Safdie's Habitat 67, not tamed and still a barrier to shipping, which now hugs the south shore of the river and passes through the concrete locks of the St. Lawrence Seaway. A little farther up the river is another, bigger set of rapids, called Lachine, a rather derisive reference to the illusions or delusions of explorers who thought that China might lie on the other side.[5] As we know now, it's half a world away, although what happens there now affects us all.

We stopped a couple of times before we got to the point where the river narrows, once getting out to stretch our legs at Sainte-Flavie, a pretty little town with small farms on its outskirts. The river is tidal here, as it is as far as Quebec City, nearly 300 kilometres to the west. Concrete met us here in the form of a dozen or more concrete sculptures—*Le Grand rassemblement*—created over several years by artist Marcel Gagnon in his basement and then dragged to the shore. From a distance the statues, which weigh about a ton each, look like they might be primitively carved logs. The features of each are sketchy, only suggesting human faces and forms. Gagnon has placed them in the

Figure 6.1: *Le Grand rassemblement* at Sainte-Flavie, where concrete sculptures meet the rising waters of the St. Lawrence estuary. Photo: Mary Soderstrom.

tidal plain on the river side of his house. Most of them gaze toward the centre of the river. At high tide they are all but submerged. So, too, will be many of the properties along this stretch of the road: in 2019 the municipality began a program to encourage people living along the waterfront to move to higher ground.[6]

And therein lies the take-home lesson. It's elementary, dear Watson: concrete and what we do with it will determine our future, as the climate changes and the waters rise, unless we figure out a way, collectively, to radically change what we build and how we build it.

APPENDIX:

CONCRETE IN TWENTY ICONIC CONSTRUCTIONS

1. **HARBOUR INSTALLATIONS, CAESARA MARITIMA, ON THE COAST OF WHAT IS NOW ISRAEL, 23–12 BCE:** These are a brilliant example of Roman concrete construction. To build them, between 23 and 12 BCE the Romans imported about 52,000 tons of volcanic sand needed to make concrete that sets in water. That was when the Middle East supplied great quantities of grain to Rome. Providing a safe harbour for this trade was a major reason for building the installations. https://web.uvic.ca/~jpoleson/ROMACONS/Caesarea2005.htm

2. **PANTHEON, ROME, 125 CE:** The dome of the Pantheon is still the largest concrete dome that isn't reinforced with steel. The excellent concrete used to form the dome is partly responsible for its longevity, but also important was what happened after the fall of Rome. The building became a Christian church in 609 CE, which meant that it was not abandoned for any length of time. In addition, the original wisdom of its architects can be seen in the way that the fissures that have developed over time seem to have had little effect

on the soundness of the building. http://nautil.us/issue/24/error/why-we-should-let-the-pantheon-crack

3. **EDDYSTONE LIGHTHOUSE, ENGLAND, COMPLETED 1759:** Using a mortar developed by John Smeaton that set in water, the mythic lighthouse was in service for 127 years before being decommissioned. Thirty years after completion, Smeaton noted that the mortar under water was holding well but that the rock above the high water line was "sensibly corroded." In the end, it was wave action undercutting the very rock on which the light stood that doomed it. https://fineartamerica.com/featured/artwork-of-the-third-eddystone-lighthouse-science-photo-library.html

4. **PONT SOUILLAC, FRANCE, COMPLETED 1824:** Louis Vicat, a young French military engineer charged with building a bridge over the Dordogne River, suspected that the key to hydraulic cement lay in compounds that were created in the heat of volcanic action. Over nearly a decade he experimented with different processes through which silica and alumina are combined with lime at great heat. He published his findings in 1817 while he waited to get funding for the Souillac bridge project. Because he didn't apply for a patent for his method, unlike other engineers and tinkerers trying to solve the puzzle of how to make hydraulic cement, his contribution has gone largely unrecognized except in France. https://www.vicat.com/about-us/vision/history-of-louis-vicat

5. **ERIE CANAL, NEW YORK STATE, UNITED STATES, COMPLETED 1825:** Before the development of the railroad, canals were the most efficient way of transporting goods. The early nineteenth century saw a great expansion of the canal network. The canal connecting the Hudson River to Lake Erie was built using "natural" cement made from rock quarried in New York State. Eclipsed by railroads within a few decades, inland waterways nevertheless became as mythic as the Eddystone Lighthouse, with songs about the Erie Canal entering into the bedrock of American culture. https://www.history.com/topics/landmarks/erie-canal

6. **I, RUE DANTON, PARIS, FRANCE, 1892:** One of the first buildings constructed completely with reinforced concrete, this structure shows the versatility of the material: Art Nouveau moulding, sculptures, and corbels were poured concrete. https://www.unjourdeplusaparis.com/en/paris-insolite/rue-danton-premier-immeuble-beton-arme-paris

7. **GRAIN ELEVATORS, NORTH AMERICA, EARLY TWENTIETH CENTURY:** Although they were sometimes constructed of steel or wood, the great cylindrical silos built of concrete influenced a generation of architects who found inspiration in their colossal mass and unornamented simplicity in the early twentieth century. Le Corbusier called them "the magnificent first fruits of the New Age." https://www.nytimes.com/1984/08/19/nyregion/buffalo-s-grain-elevators-inspiration-or-a-blight.html?searchResultPosition=7

8. **THE NARKOMFIN BUILDING, MOSCOW, USSR, 1930:** Built of reinforced concrete, the building was intended to house employees of the Commisariat of Finance (Narodnyo Kommissariat Finansov, or Narkomfin in popular parlance) and to display the very latest in architectural design and communal living. Swiss architect Le Corbusier was very much influenced by its design. Despite many changes, the building still exists, and in 2018 it began its newest incarnation as a luxury apartment house. https://archpaper.com/2016/08/narkomfin-constructivism-restoration/

9. **HIGH-RISE "TOWER IN THE PARK" HOUSING, CITIES ALL OVER THE WORLD, 1950S TO PRESENT:** Le Corbusier championed the separation of cities into segments, with residential housing concentrated in verdant parks where high-rise buildings housed everyone. The model has been used around the world with mixed results. Public housing of this type built in North America and most of Europe has not been maintained and in many cases has become high-rise slums. One place where it has worked is Singapore. https://skyrisecities.com/news/2015/10/death-highrise-brief-history-us-public-housing-policy; https://www.hdb.gov.sg/cs/infoweb/press-releases/hdb-receives-global-recognition-03112016

10. **QUEEN ELIZABETH WAY, TORONTO, CANADA'S FIRST FOUR LANE HIGHWAY, OPENED 1939:** Urban development in the twentieth century in North America depended on an ever-growing network of highways, most of which are partly constructed with concrete. Calculations based on studies made in California suggest that each additional kilometre of a traffic lane adds up to 1309.44 kg more CO_2 per day. Multiply that by 365 days a year, then figure out how many new traffic lanes there are in all, and you arrive at a very substantial number. https://www.thespec.com/news-story/6707379-the-qew-canada-s-first-super-highway-turns-77/

11. **LEVITTOWN, NEW YORK, 1947 ONWARD:** Following World War II, urban areas in North America were remade by millions of single-family houses, frequently constructed using the slab-on-grade method where a pad was bulldozed in the earth, a concrete slab laid, and the house framed and built on top of it. The Levittown projects and others like them used mass-production techniques to build millions of houses. https://www.theguardian.com/cities/2015/apr/28/levittown-america-prototypical-suburb-history-cities

12. **PISO FIRME, MEXICO, AND OTHER DO-IT-YOURSELF CONCRETE PROJECTS, TWENTIETH CENTURY:** Home-made concrete and prefab concrete blocks have improved living conditions for people all over the world. An example: Mexico's *piso firme* initiative, which subsidizes putting in concrete floors in existing housing. The impermeable floors stop parasites which live in dirt from infecting people with the result that children grow up healthier and do better in school. http://www.pa.gob.mx/publica/rev_40/NOTAS/Ernesto%20Cordero%20Arroyo.pdf

13. **MAGINOT LINE, FRANCE, COMPLETED 1940:** Following World War I, the French built more than 700 kilometres of fortifications, tunnels, and obstacles along its borders with Belgium, Luxembourg, Germany, Switzerland, and Italy. The defence was constructed between 1928 and 1940, and required millions of tons of concrete.

Despite its strength and elaborate design, the line was unable to prevent an invasion by German troops who entered France via Belgium in May 1940. https://militaryhistorynow.com/2017/05/07/the-great-wall-of-france-11-remarkable-facts-about-the-maginot-line/

14. **MISSILE SILOS, UNITED STATES, 1960S–70S:** The Cold War between the United States and the USSR was predicated on mutual total destruction, and the nuclear weapons that were to deliver that were stored deep underground in concrete silos. Neither they nor the handful of shelters intended to protect military commanders and civilian leaders would have been possible without high-density, reinforced concrete designed to withstand a megaton atomic blast. http://plainshumanities.unl.edu/encyclopedia/doc/egp.ii.042

15. **HOOVER DAM, NEVADA AND ARIZONA, UNITED STATES, CONSTRUCTION 1931–36:** This dam on the Colorado River in the United States is emblematic of all the dams built over the last eighty years to generate electricity, provide water for irrigation, and control floods. Many projects built during the period also aimed both to provide work and to build a better world. Agriculture, settlement patterns, and present-day life would be inconceivable without these massive dam projects. https://www.usbr.gov/lc/hooverdam/

16. **FRIANT-KERN CANAL, CALIFORNIA, UNITED STATES, CONSTRUCTION 1949–51:** The water of the 245-kilometre-long Friant-Kern canal, like other canal projects in the American West and the more recent South North Water Transfer Project in China, irrigates billions of dollars of fruits and vegetables. The adjacent Fresno County alone grows more agricultural crops than twenty of the fifty states in the United States. According to the most recent statistics, in 2018 it produced nearly US$1.2 billion worth of almonds, US$1.1 billion worth of grapes, US$862 million of pistachios, and more than half a billion U.S. dollars of poultry. https://www.watereducation.org/aquapedia/friant-kern-canal

17. **ASSEMBLY HALL, DHAKA, BANGLADESH, 1970:** American architect Louis I. Kahn played with light when he created the mosque that is the entrance to the legislative complex in Dacca (now Dhaka), Bangladesh. Because Pakistanis pray five times each day, Kahn said he wanted to make the Assembly Hall a "transcendent place where no matter what kind of rogue you are, when you go into the Assembly, somehow you should vote for the right thing." The result is a chamber where light streams indirectly down, in hopes that it will illuminate politics as well. https://www.archdaily.com/83071/ad-classics-national-assembly-building-of-bangladesh-louis-kahn

18. **LA GRANDE ARCHE DE LA DEFENSE, PARIS, FRANCE, OPENED 1989:** To celebrate the 200th anniversary of the French Revolution an architectural competition was held, and the winner was a hollow white cube to be made of concrete covered with marble. The dimensions were such that the Notre-Dame de Paris cathedral could fit inside the arch formed by the sides and roof of the cube. While La Grande Arche has become a Parisian icon, it also demonstrates how difficult it is to make an idea concrete. https://en.wikipedia.org/wiki/Grande_Arche#/media/File:La_Grande_Arche_de_la_D%C3%A9fense.jpg

19. **FALLINGWATER HOUSE, MILL RUN, PENNSYLVANIA, 1936:** American architect Frank Lloyd Wright designed this cantilevered house set above a waterfall in the woods of rural Pennsylvania. Built using reinforced concrete, it has been considered one of the architect's most successful buildings. Nevertheless, the original plans did not call for enough steel reinforcement, and despite engineers insisting that the building have double the reinforcement Wright called for, it has needed considerable maintenance. https://fallingwater.org/world-heritage-site/

20. **CHAMPLAIN BRIDGE, MONTREAL, CANADA, COMPLETED 1962:** The bridge was terribly designed, an example of concrete gone wrong. Its prestressed concrete structure featured elements that were stretched and then embedded in concrete poured around it. By the mid-2010s, however, the concrete and steel became so badly eroded that the bridge had to be replaced. The new Samuel de Champlain Bridge opened in 2019. https://www.newchamplain.ca/

NOTES

CHAPTER I

1   See also two of my other books: *Green City: People, Nature and Urban Places* (Montreal: Véhicule Press, 2006) and *The Walkable City: From Haussmann's Boulevards to Jane Jacobs' Streets and Beyond* (Montreal: Véhicule Press, 2008.)

2   Marie-Louise Inizan and Jacques Tixier, "L'émergence des arts du feu : le traitement thermique des roches siliceuses," in "La pyrotechnologie à ses débuts. Evolution des premières industries faisant usage du feu," ed. Andreas Hauptman, special issue, *Paléorient* 26, no. 2 (2000): 23–36, doi:10.3406/paleo.2000.4707 http://www.persee.fr/doc/paleo_0153-9345_2000_num_26_2_4707; and Gerd Weisgerbe and Lynn Willies, "The use of fire in prehistoric and ancient mining-firesetting," also in "La pyrotechnologie à ses débuts. Evolution des premières industries faisant usage du feu," ed. Andreas Hauptman, special issue, *Paléorient* 26, no. 2 (2000): 131–49, doi:10.3406/paleo.2000.4715.

3   "PRIMITIVE FIRE SCIENCE! Burning Shells into Lime in a Grass Straw & Clay Furnace," YouTube video, 6:53, posted by "SkillCult," January 22, 2016, https://www.youtube.com/watch?v=e54ISzGasdo.

4   Lawrence L. Langer, *Pre-empting the Holocaust* (New Haven, CT: Yale University Press, 1998), 21.

5   A. Hauptmann and Ü. Yalcin, "Lime Plaster, Cement and the First Puzzolanic Reaction," in "La pyrotechnologie à ses débuts. Evolution des premières industries faisant usage du feu," ed. Andreas Hauptmann, special issue, *Paléorient* 26, no. 2 (2000): 61–81, doi:10.3406/paleo.2000.4710.

6   Li Zuixiong, Zhao Linyi, Li Li, and Wang Jinua, "Research on the modification of two traditional building materials in ancient China," *Heritage Science* 1, no. 27 (2013), doi:10.1186/2050-7445-1-27.

7   Fuwei Yang, Bingjian Zhang, and Qinglin Ma, "Study of Sticky Rice–Lime Mortar Technology for the Restoration of Historical Masonry Construction," *Accounts of Chemical Research* 43, no. 6 (2010): 936–44, doi:10.1021/ar9001944.

8   Gregory L. Possehl, *The Indus Civilization: A Contemporary Perspective* (New York: AltaMira Press, 2002), 89.

9   Nick Gromicko and Kenton Shepard, "The History of Concrete," International Association of Certified Home Inspectors, https://www.nachi.org/history-of-concrete.htm. Gypsum plaster can be made by temperatures about 150 degrees C, which would be an advantage in Egypt which seems to have been much poorer in wood stocks than the Italian peninsula, even in the times of the Pharaohs.

10   Jolene K. Johnson, "Hohokam Ecology: The Ancient Desert People and their Environment," master's thesis, Arizona State University, 1997, https://archive.org/stream/hohokamecologyanoojohn/hohokamecologyanoojohn_djvu.txt, Sep 2 1999.

11   Isabel Villasen and Elizabeth Graham, "The use of volcanic materials for the manufacture of pozzolanic plasters in the Maya lowlands: a preliminary report," *Journal of Archaeological Science* 37 (2010): 1339–47, doi:10.1016/j.jas.2009.12.038.

12   Atlanta Preservation and Planning Services, "Introduction to Tabby," compiled from Lauren B. Sickels-Taves and Michael S. Sheehan, *The Lost Art of Tabby: Preserving Oglethorpe's Architectural Legacy* (Southfield, MI: Architectural Conservation Press, 1999), Atlanta Preservation, http://atlantapreservation.com/buildingmaterials/TabbyInfo.pdf.

13   Oxygen was recognized as a true element by Antoine Lavoisier in 1777, along with carbon (1789) and hydrogen (1798). Humphrey Davy discovered calcium by electrolysis of quicklime in 1808.

14   The name, by the way, is sometimes given as Ciment McInnis. That's the French moniker. This is a very Quebec endeavour underwritten by millions of dollars invested by Quebec taxpayers and the Quebec pension fund. The McInnis part of the name comes from a local family who long ago gave their name to the cove where the plant is located.

15   The idea that the universe is made up of four elements—earth, fire, water, and air—was the background world view that the Roman engineers who perfected their version of concrete worked within. The philosophical concept itself can be traced back to the Greek philosopher Empedocles in about 450 BCE. It was elaborated on eighty or so years later by Aristotle (who added a fifth element, aether, not found on Earth, but which filled the starry skies).

CHAPTER 2

1   Genivar Inc., "Projet d'implantation d'une cimenterie sur le territoire de Port-Daniel-Gascons: Mise à jour de l'étude de répercussion sur l'environnement. Document

consolidé pour diffusion publique intégrant le rapport original et les analyses supplémentaires demandées par le MDDEFP" (Montreal: GENIVAR Inc., 2013), 1.

2   There was a series of newspaper and electronic media reports about this. See, among others, Sophie Cousineau, "Quebec's new McInnis Cement project a gamble for taxpayers," *Globe and Mail*, January 31, 2014, https://www. theglobeandmail.com/report-on-business/quebecs-new-mcinnis-cement-project-a-gamble-for-taxpayers/article16646442/; also Martin Croteau, "Le plus gros pollueur du Québec pas encore soumis au marché du carbone," *La Presse*, April 23, 2018, http://www.lapresse.ca/environnement/201804/22/01-5162063-le-plus-gros-pollueur-du-quebec-pas-encore-soumis-au-marche-du-carbone.php.

3   Jean-Marie Fallu, *Une histoire d'apparentance La Gaspésie* (Québec: Les Édition GID, 2004), 312.

4   Ibid., 321.

5   Genivar Inc., "Projet d'implantation," 7, 81.

6   See also, François Normand, "La stratégie nord-américaine de McInnis Cement," *Les Affaires*, May 6, 2017, http://www.lesaffaires.com/secteurs-d-activite/general/la-strategie-nord-americaine-de-ciment-mcinnis/594690.

7   "Trinity Rail delivers new railcars to McInnis Cement," Railway Pro, April 5, 2017, http://www.railwaypro.com/wp/canada-trinityrail-delivers-new-railcars-mcinnis-cement/.

8   "Réhabilitation du chemin de fer de la Gaspésie. Le premier ministre annonce le début des travaux pour le tronçon entre Matapédia et Caplan," Transports Québec, April 29, 2019, https://www.transports.gouv.qc.ca/fr/salle-de-presse/nouvelles/Pages/annonce-debut-travaux-troncon-matapedia-caplan.aspx.

9   Genivar Inc., "Projet d'implantation," 84.

10   Gaspesia in Chandler and Smurti-Stone in New Richmond, to name only two.

11   Gilles Philibert, "Du crabe des neiges en abondance pour les pêcheurs de la Gaspésie," Radio Gaspesie, August 11, 2017, https://radiogaspesie.ca/nouvelles/actualite/crabe-neiges-abondance-pecheurs-de-gaspesie/.

12   "EI Fishing benefits – Overview," Government of Canada, July 30, 2019, https://www.canada.ca/en/services/benefits/ei/ei-fishing.html.

13   William Faulkner, *As I Lay Dying* (New York: Vintage Books, 1990), 213.

14   Ibid., 240

15   For a wonderfully clear and well-written consideration of Roman concrete and all its implications, see Lynne C. Lancaster, *Concrete Vaulted Construction in Imperial Rome: Innovations in Context* (Cambridge: Cambridge University Press, 2005). She is able to explain the chemical processes better than anyone else I've encountered, giving an understanding that resonates to the present day. The introduction and chapters on "Centering and Formwork" and "Ingredients: Mortar and Caementa" are particularly interesting and useful.

16 Marie D. Jackson, et al., "Phillipsite and Al-tobermorite mineral cements produced through low-temperature water-rock reactions in Roman marine concrete" *American Mineralogist* 102, no. 7 (2017): 1435–50, doi.org/10.2138/am-2017-5993CCBY.

17 Villasen and Graham, "The use of volcanic materials."

18 Tacitus, *The Annals of Tacitus*, Book xv, 38–44, http://penelope.uchicago.edu/Thayer/E/Roman/Texts/Tacitus/Annals/15B*.html.

19 Many of the partially excavated buildings have decorated plaster surfaces over facing stone with cemented rubble as a core, particularly in walls facing on the street. See Andrew Wallace-Handrill, *Houses and Society in Pompeii and Herculaneum* (Princeton, NJ: Princeton University Press, 1994), particularly photographs on pp. 119 and 176, as well as the official Pompei Archeological Park website, http://www.pompeiisites.org/Sezione.jsp?titolo=Pompeii+Projects&idSezione=985.

20 R.L. Hohlfelder, C. Brandon, and J.P. Oleson, "Constructing the Harbour of Cesarea on the Sea: New Evidence from the ROMACONS Field Campaign of October 2005," *The International Journal of Nautical Archaeology* 36, no. 2 (2007): 409–15, https://web.uvic.ca/~jpoleson/ROMACONS/Caesarea2005.htm.

21 Piero Gianfrotta, "Comments Concerning Recent Fieldwork on Roman Maritime Concrete," *International Journal of Nautical Archaeology* 40, no. 1 (2011): 188–93, doi:10.1111/j.1095-9270.2010.00305.

22 Marcus Vitruvius Pollio, "On Lime," Book II, Chapter V in *Ten Books on Architecture*, trans. Joseph Gwilt (London: Longman, Brown, Green, and Longmans, 1826), https://archive.org/stream/architecturemaroogwilgoog/architecturemaroogwilgoog_djvu.txt.

23 Charles J. Ferguson, "The Growth of Architecture: being the address at the opening meeting of the Architectural Section at the Shrewsbury Meeting," *F.S.A The Archeological Journal* 51 (1894): 325–36, http://archaeologydataservice.ac.uk/archives/view/archjournal/contents.cfm?vol=51&CFID=21d33278-5751-49fd-9eb8-b46bb5d28297&CFTOKEN=0.

24 Richard W. Steiger, "From prehistoric rubble mixes to Roman cement," The Concrete Producer (1995), http://www.theconcreteproducer.com/_view-object?id=00000154-1cf0-db06-a1fe-7ff8c33c0000.

25 The Society of Estate Clerks of Works, *The Journal of the Society of Estate Clerks of Works*, 8–9 (1895–1896), 84.

26 "Lowest foundation deposit of the eastern Bell Tower wall (206). Solid foundation of limestone mortar. No stonework observed though probably formed by pouring mortar slurry into foundation trench (258), reinforced by the addition of flint and other stones work to create a solid base 80 cm high and overlain by keying-in stones (251) and (243)." Wessex Archaeology, "Salisbury Cathedral, Wiltshire Archaeological Evaluation and Assessment of Results," (2009):

38, https://www.wessexarch.co.uk/sites/default/files/68741_Salisbury%20 Cathedral.pdf.

27   Robert A. Scott, *The Gothic Enterprise: A Guide to Understanding the Medieval Cathedral* (Berkeley: University of California Press, 2003), 23. Note that this mortar appears not to have been hydraulic since it took months if not years to dry.

28   "Viollet-le-Duc—Dictionnaire raisonné de l'architecture française du XIe au XVI siècle, 1854–1868, tome 4.djvu/178," Wikisource, https://fr.wikisource.org/wiki/ Page:Viollet-le-Duc_-_Dictionnaire_raisonn%C3%A9_de_l%E2%80%99architec-ture_fran%C3%A7aise_du_XIe_au_XVIe_si%C3%A8cle,_1854-1868,_tome_4. djvu/178.

29   Chandra Mukerji, *Impossible Engineering: Technology and Territoriality on the Canal du Midi* (Princeton, NJ: Princeton University Press, 2009), 208.

30   Ibid., 120–22.

31   Robert L. Day, *Pozzolans for Use in Low Cost Housing: A State of the Art Report prepared for the International Development Research Centre* (Ottawa: International Development Research Centre, 1990), https://idl-bnc-idrc.dspacedirect.org/ bitstream/handle/10625/5782/49685.pdf?sequence=1.

32   T.G. Nijland and R.P.J. van Hees, "The volcanic foundation of Dutch architecture: Use of Rhenish tuff and trass in the Netherlands in the past two millennia," *Heron* 61, no. 2: 69–98, https://repository.tudelft.nl/.../uuid%3Ad7b3412a-dba3-4bc0-85ae-e338f5733f31uuid:d7b3412a-dba3-4bc0-85ae-e338f5733f31.

33   Roberto Gargiani, *Concrete from archeology to invention, 1700–1769: The Renaissance of Pozzolana and Roman Construction Techniques*, tr. Stephen Piccolo (Laussane: EOFL Press, 2013), 100. This extremely well-documented book paints the picture of the development of cement and concrete in the eighteenth century from a decidedly Italian point of view, which is refreshing after the many Anglo-centric or Franco-centric histories. However, the book suffers from an execrable translation: one hopes that the translation is more accurate than the prose is felicitous.

34   Henry Reid, *A Practical Treatise on Concrete and How to Make It: With Observations* (London: E. & F.N. Spon, 1868), 24, https://babel.hathitrust.org/cgi/ pt?id=hvd.32044091912626&view=1up&seq=11.

35   Gargiani, *Concrete*, 325.

36   James Nicholas Douglass, "Note on the Eddystone Lighthouse," *Minutes of Proceedings of the Institution of Civil Engineers* 53 (1878): 247.

37   Roman Kozłowski, David Hughes, and Johannes Weber, "Roman Cements: Key Materials of the Built Heritage of the 19th Century," in *Materials, Technologies and Practice in Historic Heritage Structures*, eds. M. Boştenaru Dan, R. Přikryl, Á. Török (Dordrecht: Springer, 2010), 259–77, doi:10.1007/978-90-481-2684-2_14.

38 Louis Vicat, *Recherches expérimentales sur le chaux de construction, les bétons et les mortiers ordinaires* (Paris, 1817), 372–4, https://archive.org/details/re-cherchesexperioovica/page/n4/mode/2up.

39 "L'Histoire de Louis Vicat," Vicat Group, https://www.vicat.fr/a-propos-de-nous/vision/l-histoire-de-louis-vicat.

40 Honoré de Balzac, *Le Curé de village* (1874; ebook, La Bibliothèque électronique du Québec), 354, https://beq.ebooksgratuits.com/balzac/Balzac-65.pdf.

41 I.L. Znachko-Ìàvorskiï, *Egor Gerasimovich Chelidze, izobretatel' tsementa* (Tbilisi: Sabchota Sakartvelo, 1969). The title of this monograph translates as "Egor Gerasimovich Chelidze: inventor of cement."

42 Robert Courland, *Concrete Planet: The Strange and Fascinating Story of the World's Most Common Man-Made Material* (Amherst, NY: Prometheus, 2011), 184.

43 "Cement Patents," Thomas A. Edison Papers website, Rutgers University, last modified, October 28, 2016, http://edison.rutgers.edu/cemepats.htm.

44 Robert W. Lesley, *History of the Portland Cement Industry in the United States with Appendices Covering Progress of the Industry by Years* (Chicago: International Trade Press, 1924), 185, https://archive.org/details/historyofportlanoolesl.

45 David C. Burkenroad, "Jamul Cement: Speculation in the San Diego Hinterland," *The Journal of San Diego History* 25, no. 4 (1979), https://sandiegohistory.org/journal/1979/october/cement/.

46 Charles A. Newhall, "Growth of Cement Industry on Pacific Coast," *Concrete-Cement Age* 2 (April 1913): 196–8.

47 Priscilla Long, "Two small industrial communities merge to form the Town of Concrete in 1909," History Link, essay 7857, July 20, 2006, http://www.historylink.org/File/7857.

48 *Histoire de raconter: Le quartier Giffard, Arrondissement de Beauport*, Ville de Québec, 2007, https://www.ville.quebec.qc.ca/publications/patrimoine/docs/histoire_de_raconter_giffard.pdf.

49 David Ferrel, "Mountain Shifts Slowly from Stone to Cement," *Los Angeles Times*, May 9, 2002, http://articles.latimes.com/2002/may/09/news/surround-09.

50 Calculated from United States official statistical data: "Wholesale Price of Portland Cement," The National Bureau of Economic Research, http://www.nber.org/databases/macrohistory/rectdata/04/m04076b.dat; and T. Wang, "U.S. cement prices 2007–2019," Statista, February 11, 2020, https://www.statista.com/statistics/219339/us-prices-of-cement/.

51 David Perilli, "Update on South Korea," Global Cement, June 28, 2017, http://www.globalcement.com/news/item/6273-update-on-south-korea.

52 Gilles Gagné, "La Caisse de dépôt prend le contrôle de Ciment McInnis," *Le Soleil*, August 11, 2016, https://www.lesoleil.com/affaires/la-caisse-de-depot-prend-le-controle-de-ciment-mcinnis-90f1230af5fcc6622d34394e3b9baf8e.

53 S. Demis, J.G. Tapali, and V.G. Papadakis, "Plant design and economics of Rice Husk Ash exploitation as a pozzolanic material," *Waste Biomass Valor* 6

(2015): 843, https://doi.org/10.1007/s12649-015-9412-1; and Guilherme Chagas Cordeiro, Romildo Dias Toledo Filho, and Eduardo de Moraes Rego Fairbairn, "Use of ultrafine rice husk ash with high-carbon content as pozzolan in high performance concrete," *Materials and Structures* 42 (2009): 983–92, doi:10.1617/s11527-008-9437-z.

54 Chris Stern, interview with the author, February 20, 2018, and email correspondence, July 23, 2019.

55 Gerald G. Kuhn and Francis P. Shepard, *Sea Cliffs, Beaches, and Coastal Valleys of San Diego County: Some Amazing Histories and Some Horrifying Implications* (Berkeley: University of California Press, 1984), 143–54.

56 Park and Recreation Department, City of San Diego, and Merkel & Associates, Inc., "San Diego River Natural Resource Management Plan," final draft, May 2006, 109, https://nrm.dfg.ca.gov/FileHandler.ashx?DocumentID=24784.

57 Lancaster, *Concrete Vaulted Construction*, 56.

58 Ian Steadman, "Poor-Quality Chinese Concrete Could Lead to Skyscraper Collapses," *Wired* uk, March 21, 2013, https://www.wired.com/2013/03/poor-quality-chinese-concrete-could-lead-to-skyscraper-collapses/.

59 United Nations Environment Programme, "Sand, Rarer Than One Thinks," unep Global Environment Alert Service report, March 2014, 5, https://na.unep.net/geas/archive/pdfs/GEAS_Mar2014_Sand_Mining.pdf.

60 Lindsay Murdoch, "Sand Wars: Singapore's Growth Comes at the Environmental Expense of Its Neighbours," *Sydney Morning Herald*, February 26, 2016, https://www.smh.com.au/world/sand-wars-singapores-growth-comes-at-the-environmental-expense-of-its-neighbours-20160225-gn3uum.html.

61 Yushun Chen, "Construction: Limit China's Sand Mining," *Nature* 550 (2017): 457, http://dx.doi.org/10.1038/5.

62 "Sustainable Sand Mining Management Guidelines," Government of India, Ministry of Environment, Forest and Climate Change, http://envfor.nic.in/sites/default/files/Final%20Sustainable%20Sand%20Mining%20Management%20Guidelines%202016.pdf.

63 "Sand mining in rivers need to be limited and regulated – Minister," *ColomboPage*, March 10, 2017, http://www.colombopage.com/archive_17A/Mar10_1489166018CH.php.

64 Henry Fountain and Ben C. Solomon, "Melting Greenland Is Awash in Sand," *The New York Times*, July 1, 2019, https://www.nytimes.com/interactive/2019/07/01/climate/greenland-glacier-melting-sand.html.

65 James Rufus Koren, "Why Builders of Big L.A. Projects Are Making Concrete with Gravel and Sand Shipped from Canada," *Los Angeles Times*, November 4, 2017, http://www.latimes.com/business/la-fi-canadian-gravel-20171104-htmlstory.html.

66 "Manufactured sand receives stamp of excellence in Sri Lanka," CDE Global, https://www.cdeglobal.com/case-studies/ca-co-sri-lanka.

67 "La généralisation de la pierre armée," Passerelle(s), Biblothèque nationale de France, http://passerelles.bnf.fr/techniques/pantheon_paris_02.php.

68 Centre d'information sur le ciment et ses applications, "Histoire du béton: Naissance et développement, de 1818 à nos jours," *Cahier des modules de conférence pour les écoles d'architecture,* B90A, 2009, http://www.infociments.fr/telecharger/CT-B90A.pdf, a publication developed for architecture schools in France by the data source of the French cement and concrete industry.

69 Mass timber construction now allows building much taller wooden structures, and building codes in many places are being modified as a consequence. mgb ARCHITECTURE + DESIGN, Equilibrium Consulting, LMDG Ltd., and BTY Group, "The Case for Tall Wood Buildings: How Mass Timber Offers a Safe, Economical, and Environmentally Friendly Alternative for Tall Building Structures," 2012, http://cwc.ca/wp-content/uploads/publications-Tall-Wood.pdf; Kenneth Chan, "BC Building Code Will Now Allow Wood Buildings to Be Taller," Vancouver Urbanized, March 13, 2019, https://dailyhive.com/vancouver/bc-building-code-tall-wood-buildings-2019. See also chapter 5, "Air."

70 "Rue Danton, le premier immeuble en béton armé de Paris," Un jour de plus a Paris, http://www.unjourdeplusaparis.com/paris-insolite/rue-danton-premier-immeuble-beton-arme-paris.

71 "History and Heritage of Civil Engineering," American Society of Civil Engineers, February 14, 2005, https://web.archive.org/web/20050214194246/, http://www.asce.org/history/build_ingalls.swf.

72 The first patents for prestressing were awarded in the late nineteenth century, but it took decades before ways were found to tame its difficulties. The first prestressed bridge was the Walnut Lane Memorial Bridge opened in 1951. Tyson Dinges, "The history of prestressed concrete: 1888 to 1963" (master's thesis, Kansas State University, 2009), http://krex.k-state.edu/dspace/handle/2097/1439.

73 "Habitat 67," Canadian Precast Prestressed Concrete Institute, 2009, http://www.cpci.ca/en/about_us/project_month/may_2009/.

74 Explanations drawn from the Portland Cement Association's website, which provides one of the more comprehensible explanations in a world of technical descriptions. "Prestressed Concrete," Portland Cement Association, http://www.cement.org/cement-concrete-applications/products/prestressed-concrete.

CHAPTER 3

1 Courland, *Concrete Planet,* 100.

2 Michèle Benoit and Roger Gratton, *Pignon sur rue: Les quartiers de Montréal* (Montreal: Guerin éditeur, 1991), 308.

3 See the chapter "Into the Woods" in my book, *Road Through Time: The Story of Humanity on the Move* (Regina: University of Regina Press, 2017), 28–44.

4 Ibid., 78.

5   J. Donald Hughes and J.V. Thirgood, "Deforestation, Erosion, and Forest Management in Ancient Greece and Rome," *Journal of Forest History* 26, no. 2 (1982): 60–75, http://www.jstor.org/stable/4004530.

6   Maryse Tremblay, email message to author, August 6, 2019.

7   National Ready Mixed Concrete Association, "Concrete CO_2 Fact Sheet" (unpublished manuscript, 2008), NRMCA publication number 2PCO2, http://www.nrmca.org/greenconcrete/concrete%20co2%20fact%20sheet%20june%20 2008.pdf.

8   Thomas Walker, et al., "Carbon Accounting for Woody Biomass from Massachusetts (USA) Managed Forests: A Framework for Determining the Temporal Impacts of Wood Biomass Energy on Atmospheric Greenhouse Gas Levels," *Journal of Sustainable Forestry* 32, no. 1–2 (2013): 130–58, https://doi.org/10.1080/10549811.2011.652019; Chadwick Dearing Oliver, et al., "Carbon, Fossil Fuel, and Biodiversity Mitigation with Wood and Forests," *Forests, Journal of Sustainable Forestry* 33, no. 3 (2014): 248–75, doi:10.1080/10549811.2013.839386; and Warren Cornwall, "Is Wood a Green Source of Energy? Scientists Are Divided," *Science*, January 5, 2017, doi:10.1126/science.aal0574.

9   Amritha Pillay, "Petcoke imports on a rise as coal substitute," *Business Standard*, May 23, 2017, http://www.business-standard.com/article/economy-policy/petcoke-imports-on-a-rise-as-coal-substitute-117052200624_1.html.

10 Jean-Claude Cousineau, "Les dosettes K-Cup® récupérées utilisées comme combustible alternatif dans la production de ciment québécois," Éco-Énergie à Montréal, December 12, 2014, https://eco-energie-montreal.com/post/holcim-van-houtte-dosettes-keurig-k-cup/.

11 Blair Rhodes, "Judge Clears Way for Lafarge to Burn Tires in Brookfield Plant Citizens Group Requested Judicial Review of Environment Minister's Decision," CBC News, March 20, 2018, https://www.cbc.ca/news/canada/nova-scotia/judge-clears-way-for-lafarge-to-burn-tires-in-brookfield-plant-1.4584720.

12 "Ward House," ASCE Metropolitan Section, accessed February 26, 2020, http://www.ascemetsection.org/committees/history-and-heritage/landmarks/ward-house.

13 2 Peter 3:5–7 (King James Version). This is Peter talking about Noah, but there's no mention of the admonition in the Old Testament stories about Noah, his ark, and the Flood.

14 A note about things I remember in this book: the interviews I've done recently are all documented by notes and/or recordings, but I also recount things that happened long before I started keeping notes. In those instances, I've had to rely on my memory, which I've decided is quite reliable. When I've been able to test something, I remember against other sources, I've been pleased at how well the two versions jibe. A case in point is one of my first memories, of being vaccinated for smallpox in a fire station in Tacoma, Washington. I can still see a flash of a nursing sister with a white religious coif coming down a circular

staircase in the building. I thought she was an angel, but my mother told me she was just someone who was helping to make us safe. I've since learned that this was part of a mass vaccination that took place in 1946 when a soldier returning from Japan to Seattle brought home the disease: fifty-one cases were reported and sixteen deaths. (See: David A. Koplow, "That Wonderful Year: Smallpox, Genetic Engineering, and Bio-Terrorism," *Georgetown Law Faculty Publications and Other Works* 120 (2003), https://scholarship.law.georgetown. edu/facpub/120). In the case recounted here, the memories of grown-ups talking about war stand out from the general background of childhood because, I suppose, they frightened me. I don't think I'm inventing them, although the drama may have been heightened over time in my mind. And of course, there's no way to check what people may have said so long ago, since everyone involved besides me is no longer with us.

15 Samuel C. Florman, *The Civilized Engineer* (New York: St. Martin's Press, 1987), 56.

16 Clayton Donnell, *The Fortifications of Verdun 1874–1917* (Oxford: Osprey Editions, 2011), 14.

17 F.H. Gailor, "The Germans' Concrete Trenches," *The London Daily Mail*, March 24, 1915, in *New York Times Current History; The European War*, vol. 2, no. 2 (May 1915), http://www.gutenberg.org/files/15479/15479-8.txt. Gailor returned to the United States in July 1916 but appears to have been swept up by the British cause nevertheless because he is shown as being a second lieutenant in the Royal Garrison Artillery in early 1918. Afterwards he joined the American Expeditionary Force when it arrived.

18 "Militaire," Fort de Cuguret, accessed February 26, 2020, http://fort-de-cuguret. com/la-vie-du-fort/lhistoire/militaire/.

19 Jaraslava Gissübelová, "Des frontières en béton et en acier," Radio Praha, September 25, 2013, http://www.radio.cz/fr/rubrique/histoire/des-frontieres-en-beton-et-en-acie.

20 Adrian Forty, *Concrete and Culture: A Material History* (London: Reaktion Books, 2012), 63–65.

21 Valentin Graff, "Après les contrats douteux en Syrie, le cimentier Lafarge-Holcim prêt à construire le mur de Trump," *France 24*, March 3, 2017, http:// www.france24.com/fr/20170309-industrie-ciment-passe-douteux-lafarge-holcim-syrie-occupation-mur-atlantique-ei. Note that the company was rebuked by the French government for offering to build the wall between the United States and Mexico in 2017, and it has been implicated in a deal with ISIS forces in Syria that would safeguard its plant at Jalabiya.

22 "California Admission Day, September 9, 1850," California Department of Parks and Recreation, https://www.parks.ca.gov/?page_id=23856. The day is still a legal holiday in the state.

23 Erwin N. Thompson, "Historic Resources, Cabrillo National Monument, Recommendations Concerning Interpretation," *The Guns of San Diego: San Diego Harbor Defenses, 1796–1947*, ed. Howard B. Overton, National Park Service, 1991, https://www.nps.gov/articles/cabrilloww2.htm.

24 Historian Alex Wallerstein's computer simulations corroborate my memories, although I haven't been able to find the copies of the San Diego newspaper maps. See Alex Wallerstein, "Nukemap," Nuclear Secrecy, https://nuclearsecrecy.com/nukemap/, and Alex Wallerstein, "The President and the Bomb, Part IV," Restricted Data: The Nuclear Secrecy Blog, http://blog.nuclearsecrecy.com/.

25 "A Look Back at America's Fallout Shelter Fatuation," CBS News, October 7, 2010, https://www.cbsnews.com/news/a-look-back-at-americas-fallout-shelter-fatuation/.

26 "The Secret Bunker Congress Never Used," NPR, March 26, 2011, https://www.npr.org/2011/03/26/134379296/the-secret-bunker-congress-never-used.

27 Peter Kenter, "Diefenbunker construction was design-build forerunner," Journal of Commerce, November 26, 2012, https://canada.constructconnect.com/joc/news/projects/2012/11/diefenbunker-construction-was-design-build-forerunner-joc052929w; and "About the Diefenbunker," Diefenbunker Museum, 2019, https://diefenbunker.ca/about-the-diefenbunker/.

28 See the video "Preserving our Past: Stories and Structures from the Bunker," YouTube video, 6:15, posted by the Diefenbunker Museum, October 24, 2012, https://www.youtube.com/watch?v=u4ejb9_H844.

29 Daniel Howden, "Albania's relics of paranoid past," BBC News, World Edition, July 5, 2002, http://news.bbc.co.uk/2/hi/europe/2098705.stm.

30 Vladimir Isachenkov, "Vladimir Putin boasts Russia's nuclear weapons can pierce any defense," PBS, March 1, 2018, https://www.pbs.org/newshour/world/vladimir-putin-boasts-russias-nuclear-weapons-can-pierce-any-defense; and "Donald Trump in tweet vows 'end of Iran' if it wants to fight," *South China Morning Post*, May 20, 2019, article originally published in *The Guardian*, https://www.scmp.com/news/world/united-states-canada/article/3010872/never-threaten-united-states-again-donald-trump.

31 Julie Turkewitz, "A Boom Time for the Bunker Business and Doomsday Capitalists," *The New York Times*, August 13, 2019, https://www.nytimes.com/2019/08/13/us/apocalypse-doomsday-capitalists.html?fallback=0&recId=1PPvkZ4YpWt-5tR2BaCsXLoCIWwn&locked=0&geoContinent=NA&geoRegion=QC&recAlloc=top_conversion&geoCountry=CA&blockId=most-popular&imp_id=845968827&action=click&module=Most%20Popular&pgtype=Homepage.

32 "Outline History of Nuclear Energy," World Nuclear Association, February 2020, http://www.world-nuclear.org/information-library/current-and-future-generation/outline-history-of-nuclear-energy.aspx. The World Nuclear Association is an association of government and private entities engaged in making nuclear power. Canadian Nuclear Laboratories and Candu are members,

and, according to the association, it has members in 80 percent of the countries generating nuclear power.

33 Chernobyl, Three Mile Island, and the earthquake-induced disaster at Fukushima, Japan, have led many countries to turn away from using nuclear power for electricity generation. That China continues to do so makes it an outlier in the global economy.

34 See The Chernobyl Gallery, a wonderful site of archival photographs and video footage that shows the reactor and Pripyat, the town where the Chernobyl reactor's personnel lived, in detail. It is worth noting that so great was the need for electricity that Reactor 3 at Chernobyl was restarted within six months and that the facility continued operating until 2000. Note also that a small town across the current border with Ukraine in the south is also called Chernobyl. It is where tours to the Exclusion Zone leave from now. "Timeline," The Chernobyl Gallery, accessed February 26, 2020, http://chernobylgallery. com/chernobyl-disaster/timeline/.

35 Chernobyl: 51.2763 degrees N; Regina: 50.4252 degrees N.

36 Kim Willsher, "Chernobyl 30 Years On: Former Residents Remember Life in the Ghost City of Pripyat," The Guardian, March 7, 2016, https://www.theguardian. com/cities/2016/mar/07/chernobyl-30-years-residents-life-ghost-city-pripyat. See also the HBO series about the tragedy, Chernobyl (2019).

37 "Disaster at Chernobyl," Zero Hour, season 1, episode 1, directed by Renny Bartlett, aired 2004. Available as a YouTube video, 46:41, posted by "Finest," October 28, 2013, https://www.youtube.com/watch?v=ITEXGdht3y8. Zero Hour is a Canadian-British documentary-style television series which aired on The History Channel in the United States, History Television in Canada, and on the BBC in the United Kingdom.

38 Svetlana Alexievich, Voices from Chernobyl: The Oral History of a Nuclear Disaster, trans. Keith Gessen (London: Picador, 2006), 75.

39 Ibid., 194.

40 Ibid., 84.

41 "Safety Code 35: Safety Procedures for the Installation, Use and Control of x-ray Equipment in Large Medical Radiological Facilities," Health Publications, 2008, https://www.canada.ca/en/health-canada/services/environmental-work-place-health/reports-publications/radiation/safety-code-35-safety-procedures-installation-use-control-equipment-large-medical-radiological-facilities-safety-code. html.

42 Radiation-shielding concrete uses naturally heavy aggregates containing elements with high atomic weights, such as barium, instead of ordinary small stones and gravel. When using them, the concrete's density will be 3,900 kilograms per cubic metre, or 60 percent greater than normal concrete. See "Concrete @ Your Fingertips," The Concrete Society, accessed February 26, 2020, http:// www.concrete.org.uk/fingertips-nuggets.asp?cmd=display&id=786.

43 Mikhail Gorbachev, "Turning Point at Chernobyl," Project Syndicate, April 14, 2006, https://www.project-syndicate.org/commentary/turning-point-at-chernobyl?barrier=accesspaylog.

44 See Stephen Pinker, *The Better Angels of Our Nature: Why Violence Has Declined* (New York: Viking Books, 2011). It is a book that gives some hope in these troubled times.

45 John Spencer, "The Most Effective Weapon on the Modern Battlefield Is Concrete," Modern War Institute, November 14, 2016, https://mwi.usma.edu/effective-weapon-modern-battlefield-concrete/.

46 For more, see Soderstrom, *The Walkable City*.

47 Gulnaz Khan, "Thirty-one years after the worst nuclear disaster in history, a group of self-proclaimed 'stalkers' makes illegal trips into the abandoned radioactive city," *National Geographic*, December 21, 2017, https://www.nationalgeographic.com/travel/destinations/europe/ukraine/exclusion-zone-chernobyl-ukraine/.

48 "Cómo hacer la mezcla de Cemento Perfecta," YouTube video, 1:47, posted by Cemento Interoceanico, October 15, 2015, https://www.youtube.com/watch?v=mYcoooIv_1w.

49 Ernesto Cordero Arroyo, "Mejoramiento de la vivienda rural: impacto de la instalación de piso firme y estufas ecológicas en las condiciones de vida de los hogares," *Estudios Agrarios*, March 19, 2009, http://www.pa.gob.mx/publica/rev_40/NOTAS/Ernesto%20Cordero%20Arroyo.pdf.

50 Romain de Laubier, Marius Wunder, Sven Witthöft, and Christoph Rothballer, "Will 3D Printing Remodel the Construction Industry?," BCG Canada, January 23, 2018, https://www.bcg.com/en-ca/publications/2018/will-3d-printing-remodel-construction-industry.aspx. BCG Canada is a consulting firm.

51 Zigor Aldama, "'We Could 3D-print Trump's Wall': China Construction Visionaries Set to Revolutionize an Industry Rife with Graft and Old Thinking," *South China Morning Post*, May 13, 2017, https://www.scmp.com/magazines/post-magazine/long-reads/article/2093914/we-could-3d-print-trumps-wall-china-construction.

52 Peter Collins, *Concrete: The Vision of a New Architecture* (Montreal and Kingston: McGill-Queen's University Press, 2004), 41.

53 Peter Flagg Maxson and Jerry M. Sullivan, "Concrete Was Big in Seguin, Texas," *Concrete Construction Magazine*, February 1, 1991, http://www.concreteconstruction.net/business/concrete-was-big-in-seguin-texas_0.

54 "Set in Concrete: Edison Concept Houses," Indiana Landmarks, June 26, 2016, https://www.indianalandmarks.org/2016/06/set-in-concrete-edison-concept-houses/.

55 Alex Madrigal, "The Racist Housing Policy that Made Your Neighborhood," *The Atlantic*, May 22, 2014, https://www.theatlantic.com/business/archive/2014/05/the-racist-housing-policy-that-made-your-neighborhood/371439/; "1934–1968: FHA Mortgage Insurance Requirements Utilize Redlining," The Fair Housing

Center of Greater Boston, accessed February 26, 2020, http://www.bostonfair-housing.org/timeline/1934-1968-FHA-Redlining.html; and Kevin M. Kruse, "What Does a Traffic Jam in Atlanta Have to Do with Segregation? Quite a Lot," *The New York Times*, August 14, 2019, https://www.nytimes.com/interactive/2019/08/14/magazine/traffic-atlanta-segregation.html.

56   "Roman Roads," UNRV.com, accessed February 26, 2020, https://www.unrv.com/culture/roman-road-construction.php. The United Nations of Roma Victrix website is maintained by amateur historians, classicists, and archaeologists.

57   "First Concrete Pavement," American Society of Civil Engineers, accessed February 26, 2020, https://www.asce.org/project/first-concrete-pavement/.

58   Heather Brown, "Good Question: How Long Are Our Roads Supposed to Last?," CBS Minnesota, April 4, 2016, http://minnesota.cbslocal.com/2016/04/04/good-question-how-long-are-roads-supposed-to-last/.

59   "The world's biggest road networks," Road Traffic Technology, January 12, 2014, https://www.roadtraffic-technology.com/features/featurethe-worlds-biggest-road-networks-4159235/.

60   "Sources of Greenhouse Gas Emissions," United States Environmental Protection Agency, accessed February 26, 2020, https://www.epa.gov/ghgemissions/sources-greenhouse-gas-emissions.

61   John Rather, "If You're Thinking of Living in Levittown; Built All at Once, No Longer Look-Alikes," *The New York Times*, March 9, 2003, https://www.nytimes.com/2003/03/09/realestate/if-you-re-thinking-living-levittown-built-all-once-no-longer-look-alikes.html?rref=collection%2Fbyline%2Fjohn-rather&action=click&contentCollection=undefined&region=stream&module=stream_unit&version=search&contentPlacement=1&pgtype=collection.

62   Mairie de Noisy-le-Sec, "La reconstruction de Noisy-le-Sec," 2008, http://www.noisylesec.net/.

63   "The Great Concrete Architects — Le Corbusier," *Concrete Construction Magazine*, June 1, 1968, http://www.concreteconstruction.net/how-to/construction/the-great-concrete-architects-le-corbusier_o.

64   John Grindrod, *Concretopia A Journey around the Rebuilding of Postwar Britain* (Brecon: Old Street Publishing, 2014), 53.

65   Ibid., 169.

66   Athlyn Cathcart-Keays, "Moscow's Narkomfin building: Soviet blueprint for collective living," *The Guardian*, May 5, 2025, https://www.theguardian.com/cities/2015/may/05/moscow-narkomfin-soviet-collective-living-history-cities-50-buildings.

67   Ibid.

68   Katie Davies, "Luxury flats in Moscow's iconic Narkomfin building finally go on sale," *The Calvert Journal*, May 23, 2018, https://www.calvertjournal.com/articles/show/10117/luxury-flats-in-moscows-iconic-narkomfin-building-finally-go-on-sale.

69 Ingrid Campo-Ruiz, "Experimenting with Prototypes: Architectural Research in Sweden after Le Corbusier's Projects," (paper presented at Le Corbusier, 50 years later, Universitat politècnica de València, November 20, 2015), http://dx.doi.org/10.4995/LC2015.2015.893.

70 Grindrod, *Concretopia*, 53.

71 F.V. Gladkov, *Cement: A Novel*, trans. A.S. Athur and C. Ashleigh (Evanston, IL: Northwestern University Press, 1994), 312.

72 Forty, *Concrete and Culture*, 150–58.

73 Alec Luhn, "Moscow's Big Move: Is This the Biggest Urban Demolition Project Ever?," *The Guardian*, May 31, 2017, https://www.theguardian.com/cities/2017/mar/31/moscow-biggest-urban-demolition-project-khrushchevka-flats.

74 Vincze Miklós, "Photos of Everyday Life in Pripyat before the Chernobyl Disaster," Gizmodo, August 8, 2014, https://i09.gizmodo.com/photos-of-everyday-life-in-pripyat-before-the-chernobyl-1618107860.

75 Grindrod, *Concretopia*, 175.

76 Daria Litvinov, "The Great Leveler: Muscovites Unite Against Housing Demolitions," *Moscow Times*, April 27, 2017, https://www.themoscowtimes.com/2017/04/27/the-great-leveler-muscovites-unite-against-housing-demol itions-a57841.

77 Jane Jacobs, *The Death and Life of Great American Cities* (New York: Vintage, 1992), 191. For more about Jacobs, her thought, and her influence, see Soderstrom, *The Walkable City*.

78 Ibid, 208.

79 Grindrod, *Concretopia*, 162.

80 Luke Fiederer, "AD Classics: Pruitt-Igoe Housing Project / Minoru Yamasaki," *Architecture Daily*, May 15, 2017, https://www.archdaily.com/870685/ad-classics-pruitt-igoe-housing-project-minoru-yamasaki-st-louis-usa-modernism.

81 Adam Forrest, "Vienna's Affordable Housing Paradise: Public Housing Is the Accommodation of Last Resort in the U.S. Not So in Austria's Capital City," Huffington Post, July 18, 2018, https://www.huffingtonpost.com/entry/vienna-affordable-housing-paradise_us_5b4e0b12e4b0b15aba88c7b0?utm_source=esp&utm_medium=Email&utm_campaign=The+Cityscape&utm_term=282269&subid=26647851&CMP=cityscape.

82 "AD Interviews: Moshe Safdie," YouTube video, 6:50, posted by Architecture Daily, January 12, 2015, https://www.youtube.com/watch?v=2ZiiMr9GG64.

83 Moshe Safdie, "Habitat Original Proposal," item finding aid description, The Moshe Safdie Archive, McGill University Library, http://cac.mcgill.ca/moshesafdie/fullrecord.php?ID=11059&d=1.

84 Moshe Safde, interview by Anna Winston, *Dezeen*, December 18, 2014, https://www.dezeen.com/2014/12/18/moshe-safdie-architects-interview-movie-marina-bay-sands-development-singapore/.

85 Safdie, "Habitat Original Proposal."

86 Listing for 2600 Av. Pierre-Dupuy, apartment 232, Sutton, Quebec, accessed February 26, 2020, https://www.suttonquebec.com/en/inscription/apartment-for-sale-2600-av-pierre-dupuy-app-641-ville-marie-montreal-.html?noInscription=16054064&typeInscription=1l.

87 David Owen, "Concrete Jungle," *The New Yorker*, November 10, 2003, 62 ff.

88 Flavie Halais, "How Toronto is revitalizing its aging suburban residential towers," Citiscope, August 13, 2015, http://citiscope.org/story/2015/how-toronto-revitalizing-its-aging-suburban-residential-towers.

89 United Way Toronto, "Poverty by Postal Code 2: Vertical Poverty—Declining Income, Housing Quality and Community Life in Toronto's Inner Suburban High-Rise Apartments," 2001, https://www.unitedwaygt.org/document.doc?id=89.

90 Martin Perry, Lily Kong, and Brenda Yeoh, *Singapore: A Developmental City State* (New York: John Wiley and Sons, 1997), 48.

91 "Remaking Our Heartland," Housing and Development Board, Singapore, accessed March 27, 2020, https://www20.hdb.gov.sg/fi10/fi10349p.nsf/hdbroh/index.html; and Melissal Lin, "Toa Payoh to Get New Flats, New Parks in Makeover," *The Sunday Times*, April 23, 2017, republished on If Only Singaporeans Stopped to Think, https://ifonlysingaporeans.blogspot.com/2017/04/toa-payoh-to-get-new-flats-new-parks-in.html.

92 Contributions from both employers and employees go into three accounts which are used to finance an individual's housing, retirement, and health care needs. "What is the Central Provident Fund (CPF)," Ministry of Manpower, accessed February 26, 2020, https://www.mom.gov.sg/employment-practices/central-provident-fund/what-is-cpf.

93 Lee Kuan Yew, *From Third World to First: The Singapore Story: 1965–2000* (New York: HarperCollins, 2000), 97.

94 Wade Shepard, *Ghost Cities of China: The Story of Cities without People in the World's Most Populated Country* (London: Zed Books, 2015), 117.

95 UN Habitat, *Small-Scale Production of Portland Cement*, (UN Habitat, 1993), 5. At one point there were 200 kilns that produced 50,000 to 100,000 tons of cement a year, a figure to compare with McInnis Cement's plan to produce 2.2 million tons a year.

96 Similar housing is called called *hutangs* in Beijing. Both types have much in common with Singapore's old shophouses. Echo Zhao, "Shanghai's Incredible Longtang Alleyways," The World of Chinese, May 17, 2010, https://www.theworldofchinese.com/2010/05/shanghai-longtangs/.

97 Shepard, *Ghost Cities of China*, 123.

98 *Encyclopaedia Britanica*, s.v., "Qinhuangdao," accessed April 30, 2020, https://www.britannica.com/place/Qinhuangdao.

99 Monica Tan, "Bad to worse: ranking 74 Chinese cities by air pollution," Greenpeace, February 19, 2014, https://www.greenpeace.org/eastasia/blog/1820/bad-to-worse-ranking-74-chinese-cities-by-air-pollution/.

100 "China Home Ownership Rate," Trending Economics, accessed April 30, 2020, https://tradingeconomics.com/china/home-ownership-rate.

101 Melissa Chan, "From Boom Town to Ghost Town," *Al Jazeera*, September 11, 2011, https://www.aljazeera.com/blogs/asia/2011/09/70656.html.

102 Wade Shepard, "China's Most Infamous 'Ghost City' Is Rising from the Desert," *Forbes*, June 30, 2017, https://www.forbes.com/sites/wadeshepard/2017/06/30/ordos-chinas-most-infamous-ex-ghost-city-continues-rising/#651170e46877.

103 Shaun Fynn and Maristella Casciato, *Chandigarh Revealed: Le Corbusier's City Today* (New York: Princeton Architectural Press, 2017), 8; and Nicholas Fox Weber, *Le Corbusier: A Life* (New York: Alfred A. Knopf, 2008).

104 M.N. Sharma, chief architect of Chandigarh from 1965 to 1979, interview by Shaun Fynn, in Fynn and Casciato, *Chandigarh Revealed*, 232.

105 Iain Jackson and Jessica Holland, *The Architecture of Edwin Maxwell Fry and Jane Drew: Twentieth Century Architecture, Pioneer Modernism and the Tropics* (Abingdon: Routledge, 2014), 231.

106 Lisa D. Schrenk, "Pilotis and Sunshades: The Influence of Le Corbusier on Modern Architecture in Brazil," Curriculum Project 2007 Fulbright-Hays Seminar Abroad curriculum document, Latin American Network Information Center, accessed February 26, 2020, http://lanic.utexas.edu/project/etext/llilas/outreach/fulbright07/schrenk.pdf.

107 Will Housh, "Air Conditioning Then and Now," HVAC.com, November 30, 2019, https://www.hvac.com/blog/air-conditioning-then-and-now.

108 Fynn and Casciato, *Chandigarh Revealed*, 228–29.

109 James Crabtree, "Le Corbusier's Chandigarh: an Indian city unlike any other," *Financial Times*, July 3, 2015, https://www.ft.com/content/2a194cb4-1a8d-11e5-a130-2e7db721f996.

110 Bill Gates, "A Stunning Statistic about China and Concrete," Gates Notes, June 25, 2014, https://www.gatesnotes.com/About-Bill-Gates/Concrete-in-China; and "Consumption of cement in China from 2006 to 2016," Statista, accessed February 26, 2020, https://www.statista.com/statistics/269321/consumption-of-cement-in-china-since-2004/.

111 *The Guardian* ran a series of stories on these new cities in July 2019. They are must-reads for anyone interested in the shape of the world to come. See particularly Wade Shepard, "Should We Build New Cities from Scratch," *The Guardian*, July 10, 2019, https://www.theguardian.com/cities/2019/jul/10/should-we-build-cities-from-scratch?utm_source=Sailthru&utm_medium=email&utm_campaign=Issue:%202019-07-11%20Smart%20Cities%20Dive%20Newsletter%20%5Bissue:21849%5D&utm_term=Smart%20Cities%20Dive.

112 "China prohibits expansion of glass, cement capacity in 2018," *South China Morning Post*, February 12, 2018, http://www.scmp.com/news/china/economy/article/2132983/china-prohibits-expansion-glass-cement-capacity-2018.

113 *Encyclopaedia Britannica*, s.v. "Vesta," accessed February 26, 2020, https://www.britannica.com/topic/Vesta-Roman-goddess.

CHAPTER 4

1   P.C. Smith, "Chaleur Bay," in *The Canadian Encyclopedia*, last modified October 17, 2014, http://www.thecanadianencyclopedia.ca/en/article/chaleur-bay/.

2   The Central Valley is composed of two branches: the northern one is the Sacramento Valley and the southern one is the San Joaquin Valley.

3   Robert Autobee, *Friant Division Central Valley Project* (Washington, DC: U.S. Bureau of Reclamation Historical Review, 1994), 2.

4   Michael D. Orzolek, William J. Lamont Jr., Lynn F. Kime, and Jayson K. Harper, "Broccoli Production: Initial Investment in Broccoli Production Is Relatively Low and Can Be Throughout the Summer and Early Fall," PenState Extension, June 20, 2005, https://extension.psu.edu/broccoli-production; and Michelle LeStrange, Keith S. Mayberry, Steve T. Koike, and Jesús Valencia, "Broccoli Production in California," University of California, Division of Agriculture and Natural Resources, Publication 7211, accessed March 2, 2020, https://anrcatalog.ucanr.edu/pdf/7211.pdf.

5   Steven H. Kosmatka and Michelle L. Wilson, "Mixing Water for Concrete," in *Design and Control of Concrete Mixtures*, 16th ed. (Skokie, IL: Portland Cement Association, 2016), 171–77.

6   Interview with individual who went on to become a medical doctor but at the time was doing preparatory science courses, February 28, 2018, about an Expo '67 site (not the Habitat 67 one).

7   David A. Kuemmel and Rashad M. Hanbali, "Accident Analysis of Ice Control Operations: Final Report" (Alexandria, VA: The Salt Institute, 1992), https://epublications.marquette.edu/transportation_trc-ice/1/.

8   Joseph Stroberg, "What Happens to All the Salt We Dump on the Roads," *Smithsonian Magazine*, January 6, 2014, https://www.smithsonianmag.com/science-nature/what-happens-to-all-the-salt-we-dump-on-the-roads-180948079/.

9   "Samuel De Champlain Bridge—Frequently Asked Questions," Infrastructure Canada, August 27, 2019, http://www.infrastructure.gc.ca/nbsl-npsl/faq-eng.html.

10   "Montreal Bridge Gets $158M in Repair Funds," CBC News, March 18, 2011, https://www.cbc.ca/news/canada/montreal/montreal-bridge-gets-158m-in-repair-funds-1.1091614.

11   Muhammad Shafqat Ali, Muhammad Sagheer Aslam, and M. Saeed Mirza, "A Sustainability Assessment Framework for Bridges—A Case Study: Victoria

and Champlain Bridges, Montreal," *Structure and InfraStructure Engineering* 12, no. 11 (2016): 1381–94 http://dx.doi.org/10.1080/15732479.2015.1120754.

12   "Advancing Sustainable Materials Management: 2014 Fact Sheet Assessing Trends in Material Generation, Recycling, Composting, Combustion with Energy Recovery and Landfilling in the United States," Environmental Protection Agency, November 2016, https://www.epa.gov/sites/production/files/201611/documents/2014_smmfactsheet_508.pdf.

13   "New Turcot Will Be Built with Recycled Materials from Old One," CBC News, August 3, 2016, https://www.cbc.ca/news/canada/montreal/turcot-interchange-recycled-1.3706701.

14   "Champlain Bridge Sector: Prefeasibility study on the deconstruction of the Champlain Bridge," Ponts Jacques Cartier and Champlain Bridges, accessed March 2, 2020, http://jacquescartierchamplain.ca/community-heritage/structures-and-projects/champlain-bridge/preliminary-study-on-the-feasibility-of-the-deconstruction-of-the-champlain-bridge/?lang=en.

15   Guglielmo Mattioli, "What Caused the Genoa Bridge Collapse – and the End of an Italian National Myth?," *The Guardian*, February 28, 2019, https://www.theguardian.com/cities/2019/feb/26/what-caused-the-genoa-morandi-bridge-collapse-and-the-end-of-an-italian-national-myth.

16   Joan Didion, *The White Album* (New York: Simon and Schuster, 1979), 59.

17   Bradley J. Fikes, "Why the Ocean Is Warm, then Cold," *San Diego Union Tribune*, August 17, 2003, http://www.sandiegouniontribune.com/sdut-the-current-situation-why-the-ocean-is-warm-then-2003aug17-story.html; and the current situation: https://earthobservatory.nasa.gov/IOTD/view.php?id=87575.

18   Fresno County Annual Crop & Livestock Report, 2018, https://www.co.fresno.ca.us/Home/ShowDocument?id=37986.

19   Wallace Stegner, *Angle of Repose* (New York: Doubleday, 1971), 434.

20   The desire to tame rivers like the Colorado and the San Joaquin set the stage for the rivalry between the two dam building agencies, the Army Corps of Engineers, which focused on flood control, and the Bureau of Reclamation, which featured irrigation.

21   Didion, *The White Album*, 64.

22   Edward Charles Murphy et al., *Destructive Floods in the United States in 1905 with a Discussion of Flood Discharge and Frequency and an Index to Flood Literature* (Washington, DC: Washington Government Printing Office, 1906), 63, https://pubs.usgs.gov/wsp/0162/report.pdf.

23   David Owen, *Where the Water Goes: Life and Death along the Colorado River* (New York: Riverhead Books, 2017), 21.

24   "The Big Thompson Floor of 1976," *The Denver Post*, July 31, 2012, http://blogs.denverpost.com/library/2012/07/31/big-thompson-flood-disaster-colorado-1976/2795/.

25 See Soderstrom, *Road Through Time*, 151.

26 Alex Silveira, "O que precisa ser feito para acabar com as enchentes em Curítiba: Solução para enchentes em Curítiba passa por planejamento em relação aos cursos dos rios e 'contornar' a crescente impermeabilização do solo," *Gazeta do Povo*, February 22, 2019, https://www.tribunapr.com.br/noticias/curitiba-regiao/o-que-precisa-ser-feito-acabar-enchentes-curitiba/.

27 J. David Rogers, "Evolution of the Levee System Along the Lower Mississippi," J. Davide Rogers, personal webpage, Missouri University of Science and Technology, accessed March 3, 2020, https://web.mst.edu/~rogersda/levees/Evolution%20of%20the%20Levee%20System%20Along%20the%20Mississippi.pdf.

28 Erika Bolstad, "Irony: Levees Could Make River Flooding Worse," *Scientific American*, May 9, 2017, https://www.scientificamerican.com/article/irony-levees-could-make-river-flooding-worse/.

29 Nadja Popovich, "From Heat Waves to Hurricanes: What We Know about Extreme Weather and Climate Change," *The New York Times*, September 14, 2017, https://www.nytimes.com/interactive/2017/09/15/climate/does-climate-change-cause-hurricanes-drought.html.

30 Vivian Forbes, "Yellow River Changing Course," China Water Risk, November 12, 2015, http://chinawaterrisk.org/opinions/yellow-river-changing-course/.

31 Arindam Dey, "Failure Analysis of Sadd-el-Kafara: The Oldest Masonry Dam" (conference paper, Indian Geotechnical Conference, Kakinada, India, January 2014), https://www.researchgate.net/publication/281373650.

32 Akanksha Gupta, "The World's Oldest Dams Still in Use," Water Technology, October 20, 2013, https://www.water-technology.net/author/akanksha-guptaprogressivemediagroup-in/.

33 Donald R. Hill, "Science and Technology in Islamic Building Construction," in *Technology, Tradition and Survival: Aspects of Material Culture in the Middle East and Central Asia*, eds. Richard Tapper and Keith McLachlan (Abingdon: Routledge, 2002), 55–57.

34 "Hydroelectric Generating Stations (as at January 1st, 2019)," Hydro Québec, January 1, 2019, http://www.hydroquebec.com/generation/centrale-hydroelectrique.html.

35 Rongchao Li, "Flood Control in the Yellow River Basin in China," *Water Encyclopedia*, April 15, 2005, https://doi.org/10.1002/047147844X.sw259.

36 Conversation with docent, Hoover Dam tour, January 20, 2018.

37 Paul C. Pitzer, *Grand Coulee: Harnessing a Dream* (Pullman, WA: Washington State University Press, 1994), 212. Hoover Dam figures taken from "Hoover Dam: Frequently Asked Questions and Answers," Bureau of Reclamation, U.S. Department of the Interior, accessed March 3, 2020, https://www.usbr.gov/lc/hooverdam/faqs/damfaqs.html.

38   John Steinbeck, *The Grapes of Wrath* (New York: Viking Press, 1958), 160.

39   Ibid., 403.

40   Matthew Yglesias, "How Mass Migration Cushioned the Depression," *Slate*, April 3, 2004, http://www.slate.com/blogs/moneybox/2012/04/03/depression_migration_map_.html.

41   Pitzer, *Grand Coulee*, 196.

42   Jeff Brady, "Woody Guthrie's Fertile Month on the Columbia River," NPR, July 13, 2007, https://www.npr.org/templates/story/story.php?storyId=11918998.

43   Geoffrey C. Nunes, "Kaiser, Garfield, and Permanente," *Archives of Surgery* 137, no. 9 (2002): 1034–36, https://jamanetwork.com/journals/jamasurgery/fullarticle/212894.

44   Steinbeck, *The Grapes of Wrath*, 317.

45   Pitzer, *Grand Coulee*, 364.

46   Grapes for wine were grown in Walla Walla from the beginning of the twentieth century, but mostly by Italian immigrants for making their own wine. Then in the 1970s more serious winemaking began. The result is some wonderful wine. See also Liza B. Zimmerman, "A Closer Look at Washington's Walla Walla Wine Region," *Forbes*, July 12, 2018, https://www.forbes.com/sites/lizazimmerman/2018/07/12/the-walla-walla-travel-guide/#1ebe397130e7.

47   Virginia L. Butler and Jim E. O'Connor, "9000 Years of Salmon Fishing on the Columbia River, North America," *Quaternary Research* 62 (2004): 1–8, doi:10.1016/j.yqres.2004.03.002.

48   "Hoover Dam: Frequently Asked Questions," and "Grand Coulee Dam Statistics and Facts," *Reclamation: Managing Water in the West*, fact sheet, Bureau of Reclamation, U.S. Department of the Interior, February 2019, https://www.usbr.gov/pn/grandcoulee/pubs/factsheet.pdf.

49   "Releases: Columbia River," Safe as Mother's Milk: The Hanford Project, accessed March 3, 2020, http://www.hanfordproject.com/columbia.html.

50   "Columbia Snake River Facts," Pacific Northwest Waterways Association, www.pnwa.net/factsheets/CSRS.pdf.

51   Linda V. Mapes, "Judge: Salmon Recovery Requires Big Dam Changes on Snake River," *Seattle Times*, May 4, 2016, https://www.seattletimes.com/seattle-news/environment/lower-snake-river-dam-removal-back-on-table/.

52   "The Lower Snake River Dams Power Replacement Study," Northwest Energy Coalition, April 27, 2018, LSRDS-study-4-page-overview.pdf.

53   Jacques Leslie, "On the Northwest's Snake River, the Case for Dam Removal Grows," *Yale Environment 360*, October 10, 2019, https://e360.yale.edu/features/on-the-northwests-snake-river-the-case-for-dam-removal-grows.

54   Eric Barker, "Salmon Deal to Add More Spillage at Region's Dams," *The Lewiston Tribune*, December 20, 2018, https://www.spokesman.com/stories/2018/dec/20/salmon-deal-to-add-more-spillage-at-regions-dams/.

55 Laura L. Myers, "Salmon Revival in Sight as Elwha River Dams Fall in U.S. Northwest," Reuters, April 28, 2012, https://www.reuters.com/article/us-usa-river-washington/salmon-revival-in-sight-as-elwha-river-dams-fall-in-u-s-northwest-idUSBRE83R0DX20120428.

56 Jeff Spross, "Los Angeles Aims to Be Coal-Free in 12 Years," Think Progress, March 5, 2013, https://thinkprogress.org/los-angeles-aims-to-be-coal-free-in-12-years-12b611e530b4/.

57 Ivan Penn, "The $3 Billion Plan to Turn Hoover Dam into a Giant Battery," *The New York Times*, July 24, 2018, https://www.nytimes.com/interactive/2018/07/24/business/energy-environment/hoover-dam-renewable-energy.html.

58 Jean-Paul Vanasse, "Tout cela est bien plus que chanson," *Liberté* 10, no. 1 (January–February 1968): 6–15, https://www.erudit.org/fr/revues/liberte/1968-v10-n1-liberte1027669/29581ac.pdf.

59 "Power Transmission in Québec," Hydro Québec, accessed March 3, 2020, http://www.hydroquebec.com/learning/transport/grandes-distances.html.

60 In China lines carrying 800kv DC and 1000 kv AC have been built. See Wei Peng, "China's Silver Bullet: Can the Transmission Grid Solve China's Problems?," New Security Beat, October 20, 2017, https://www.newsecuritybeat.org/2017/10/chinas-silver-bullet-transmission-grid-solve-chinas-problems/. India is experimenting with 1200 kv AC. See M. Ramesh, "India to Have World's Highest Power Transmission Line," The Hindu Business Line, December 4, 2013, https://www.thehindubusinessline.com/economy/india-to-have-worlds-highest-power-transmission-line/article23123750.ece.

61 Tom Fennario, "James Bay Cree Fear Declining Caribou Herd," APTN News, July 31, 2015, https://aptnnews.ca/2015/07/31/james-bay-cree-fear-declining-caribou-herd/.

62 This is not to say that other hydro projects in Canada didn't damage Indigenous lands and ways of life. Flooding and mercury contamination have greatly affected Indigenous people in Manitoba, for example: James B. Waldram, "Native People and Hydroelectric Development in Northern Manitoba, 1957–1987: The Promise and the Reality," *Manitoba History* 15 (Spring, 1988), http://www.mhs.mb.ca/docs/mb_history/15/hydroelectricdevelopment.shtml.

63 Morgan Lowrie, "New York officials tour Quebec Indigenous communities to assess impact of potential hydro deal," *Globe and Mail*, July 30, 2019, https://www.theglobeandmail.com/canada/article-new-york-officials-tour-quebec-indigenous-communities-to-assess-impact/.

64 "Hydro-Québec Production," Hydro Québec, accessed March 3, 2020, http://www.hydroquebec.com/generation/.

65 Marek Warszawski, "Temperance Flat Dam Is Dead. Now Valley Lawmakers Need to Come Up with Fresh Ideas," *Fresno Bee*, May 10, 2018, http://www.fresnobee.com/opinion/opn-columns-blogs/marek-warszawski/article210649724.html#storylink=cpy.

66   Aldo Leopold, *A Sand County Almanac* (New York: Ballantine Books, 1970), 150–51.

67   John Letey, "Soil Salinity Poses Challenges for Sustainable Agriculture and Wildlife," *California Agriculture* 54, no. 2 (2000): 43–48, https://doi.org/10.3733/ca.v054n02p43.

68   Steinbeck, *The Grapes of Wrath*, 268–69.

69   Ibid., 284.

70   Philip Micklin, "The Aral Sea Disaster," *Annual Review of Earth and Planetary Sciences* 35, no. 1 (2007): 47–72, doi:10.1146/annurev.earth.35.031306.140120.

71   Liu Xianzhao and Qi Shanzhong, "Wetlands Environmental Degradation in the Yellow River Delta, Shandong Province of China," *Procedia Environmental Sciences* 11 (2011): 701–5, doi:10.1016/j.proenv.2011.12.109.

72   Jonathan Watts, "Provincial Tug-of-war Waters Down China's Yellow River Success Story," *The Guardian*, June 28, 2011, https://www.theguardian.com/environment/2011/jun/28/water-yellow-river-china; "Yellow River," Facts and Details, accessed March 3, 2020, http://factsanddetails.com/china/cat15/sub103/item448.html; and Li Yang, "Yellow River Delta Faces New Challenges," *China Daily*, April 16, 2019, http://www.chinadaily.com.cn/a/201904/16/WS5cb513d9a3104842260b6576.html.

73   Stuart Heaver, "Chairman Mao's historic swim—glorified in China but ridiculed by the rest of the world," *South China Morning Post*, August 4, 2016, http://www.scmp.com/magazines/post-magazine/long-reads/article/1999098/chairman-maos-historic-swim-glorified-china.

74   "China," International Rivers, accessed March 3, 2020, https://www.internationalrivers.org/programs/china.

75   Shepard, *Ghost Cities of China*, 162.

76   Bo Li, Songqiao Yao, Yin Yu, and Qiaoyu Guo, "The 'Last Report' on China's Rivers, Executive Summary," English translation, International Rivers, March 2014, 3, https://www.internationalrivers.org/sites/default/files/attached-files/final_rivers_report_english_small.pdf.

77   Tom Fawthrop, "Leaked Report Warns Cambodia's Biggest Dam Could 'Literally Kill' Mekong River," *The Guardian*, May 18, 2018, https://www.theguardian.com/environment/2018/may/16/leaked-report-warns-cambodias-biggest-dam-could-literally-kill-mekong-river.

78   "Dam Collapse in Laos Displaces Thousands, Exposes Dam Safety Issues," International Rivers, July 23, 2018, https://www.internationalrivers.org/dam-collapse-in-laos-displaces-thousands-exposes-dam-safety-risks.

79   Philip Ball, "The Chinese Are Obsessed with Building Big Dams," BBC, October 15, 2015, http://www.bbc.com/future/story/20151014-the-chinese-are-obsessed-with-building-giant-dams.

80   Shephard, *Ghost Cities of China*, 129.

81 Coco Liu, "Water Demands of Coal-Fired Power Drying Up Northern China," Scientific American, reprinted from Climatewire, March 25, 2013, https://www.scientificamerican.com/article/water-demands-of-coal-fired-power-drying-up-northern-china/.

82 "China," Hydro Power, May 2019, https://www.hydropower.org/country-profiles/china.

83 Ibid.

84 Lily Kuo, "China Has Launched the Largest Water-Pipeline Project in History But Is the South-to-North Water Diversion Project Creating More Problems than It Solves?," The Atlantic, March 7, 2014, https://www.theatlantic.com/international/archive/2014/03/china-has-launched-the-largest-water-pipeline-project-in-history/284300/?utm_source=twb.

85 Quoted in Kuo, "China Has Launched the Largest Water-Pipeline Project in History."

86 "Hu Jintao," People's Daily, accessed March 3, 2020, http://en.people.cn/data/people/hujintao.shtml.

87 "History," San Diego County Water Authority, accessed March 3, 2020, https://www.sdcwa.org/history.

88 Rebecca Harrington, "See How One of the World's Greatest Engineering Feats Was Created in 1915 to Bring Water to New York City," Business Insider, January 8, 2016, http://www.businessinsider.com/photos-catskill-aqueduct-nyc-incredible-engineering-2016-1.

89 "California State Water Project at a Glance," California Department of Water Resources, accessed March 3, 2020, https://water.ca.gov/LegacyFiles/recreation/brochures/pdf/swp_glance.pdf; and Alexis C. Madrigal, "American Aqueduct: The Great California Water Saga," The Atlantic, February 24, 2014, https://www.theatlantic.com/technology/archive/2014/02/american-aqueduct-the-great-california-water-saga/284009/.

90 Calculated from "Romaine lettuce," The World's Healthiest Foods, accessed March 3, 2020, http://www.whfoods.com/genpage.php?tname=foodspice&dbid=61.

91 Julia Jacobs, "Officials Identify a Source in the Romaine Lettuce E. Coli Outbreak," The New York Times, July 1, 2018, https://www.nytimes.com/2018/07/01/us/romaine-lettuce-e-coli-nyt.html?rref=collection%2Ftimestopic%2FCenters%20for%20Disease%20Control%20and%20Prevention&action=click&contentCollection=timestopics&region=stream&module=stream_unit&version=latest&contentPlacement=4&pgtype=collection.

92 Marcel Boyer, "L'exportation d'eau douce pour le développement de l'or bleu québécois," Les Cahiers de recherche de l'Institut économique de Montréal (Montreal, Institut économique de Montréal, August 2008), http://www.crecq.qc.ca/pdf/IEM-marcel-boyer.pdf.

93 Owen, *Where the Water Goes*, 234.

94 Mervyn Piesse, "Southern Water for North China: Is Water Conveyance Infrastructure a Long-Term Solution to Water Stress?," Future Directions, March 20, 2018, http://www.futuredirections.org.au/publication/southern-water-north-china-water-conveyance-infrastructure-long-term-solution-water-stress/.

95 "Frequently Asked Questions," Recharge Fresno, accessed March 3, 2020, http://www.rechargefresno.com/faqs/; and Rosanna Xia, "As Its Water Dwindled, Fresno Cracked Down Hard," *Los Angeles Times*, June 22, 2015, http://www.latimes.com/local/california/la-me-fresno-water-penalties-20150622-story.html.

96 Paul Rogers, "Drought or No Drought: Jerry Brown Sets Permanent Water Conservation Rules for Californians," *The Mercury News*, May 31, 2018, https://www.mercurynews.com/2018/05/31/california-drought-jerry-brown-sets-permanent-water-conservation-rules-with-new-laws/.

97 "Drought's Over, Yet Californians Keep Saving Water," KQED, September 6, 2017, https://www.kqed.org/science/1915327/drought-rules-lone-gone-yet-california-keeps-saving-water.

98 Owen, *Where the Water Goes*, 250.

99 "Migratory Bird Sanctuaries along the Coast of Yellow Sea-Bohai Gulf of China (Phase I)," UNESCO, accessed March 3, 2020, https://whc.unesco.org/en/list/1606/; and "Two natural sites, one in China, another in Iran, inscribed on UNESCO's World Heritage List," UNESCO, July 5, 2019, https://whc.unesco.org/en/news/2001/.

100 Gian Volpicelli, "China: The Dizzying Scale of the World's Largest Coal Hub," *Wired*, April 6, 2016, http://www.wired.co.uk/article/big-picture-china-coal-hub-power-qinhuangdao.

101 Until January 2020 the Sasaki corporate website contained a page on the Beidhaihe project, but it now seems to have been removed for reasons that are unclear. Nevertheless, images from the plan can be found elsewhere, including https://www.pinterest.ca/pin/488851734536792217/, accessed March 4, 2020.

102 Austin Williams, "Seashore Library in Beidaihe, China by Vector Architects," *Architectural Review*, November 23, 2015, https://www.architectural-review.com/buildings/library/seashore-library-in-beidaihe-china-by-vector-architects/8691051.article.

103 More than 600 Chinese cities are classified according to a tier system, purportedly to aid foreign businesses in establishing market strategies and where to invest. Dorcas Wong, "China's City-Tier Classification: How Does It Work?," China Briefing, February 27, 2019, https://www.china-briefing.com/news/chinas-city-tier-classification-defined/

104 "Firm Launches Phase II of Habitat Golden Dream Bay in Qinhuangdao," Resincow and Associates, March 22, 2018, https://resnicow.com/client-news/firm-launches-phase-ii-habitat-golden-dream-bay-qinhuangdao-china.

105   Maria Abi-Habib, "How China Got Sri Lanka to Cough Up a Port," *The New York Times*, June 25, 2018, https://www.nytimes.com/2018/06/25/world/asia/china-sri-lanka-port.html.

106   "Quick Take: Hebei to Rid Qinhuangdao of Coal in Bet on Tourism," Power Links, posted by Caixin Global, March 6, 2018, http://powerlinks.news/article/63f55f/quick-hebei-to-rid-qinhuangdao-of-coal-in-bet-on-tourism?-fieldname=industryids&t=Coal+Mining+&query=&fieldvalue=79.

107   Chris Tuckley and Adam Wu, "'Botched' Repair to China's Great Wall Provokes Outrage," *The New York Times*, September 22, 2016, https://www.nytimes.com/2016/09/23/world/asia/china-great-wall-botched-repair.html.

108   Andrew Kileen, "Cement and Sticky Rice: Where Is the Great Wall?," *The Beijinger*, September 29, 2016, https://www.thebeijinger.com/blog/2016/09/28/cement-and-sticky-rice-where-real-great-wall.

109   Spelling of names changed to reflect new orthographical conventions. "Peitaiho," Marxists Internet Archives, accessed March 3, 2020, https://www.marxists.org/reference/archive/mao/selected-works/poems/poems22.htm.

CHAPTER 5

1   Rick Scezcy, "Concrete 101," World of Concrete convention, Las Vegas, Nevada, January 2018.

2   To see what it looks like, consult one of the many videos on YouTube, such as "Watch Soap Bubbles Freeze in Real Time," YouTube video, 0:45, posted by CNN, January 16, 2017, https://www.youtube.com/watch?v=52xFz1Cn8E8, or do it yourself some cold winter day in Montreal.

3   Patrick H. Torrans and Don L. Ivey, "Review of the Literature on Air-Entrained Concrete," Research Report Number 103-1, Texas Transportation Institute, Texas A&M University, College Station, TX, February 1968, https://static.tti.tamu.edu/tti.tamu.edu/documents/103-1.pdf.

4   Lancaster, *Concrete Vaulted Construction*, 68–72.

5   "Weather and Climate Basics: What's in the Air," University Corporation for Atmospheric Research, accessed March 3, 2020, http://www.eo.ucar.edu/basics/wx_1_b_1.html.

6   Daniel H. Rothman, "Atmospheric Carbon Dioxide Levels for the Last 500 Million Years," *Proceedings of the National Academy of Sciences* 99, no. 7 (2002): 4167–71, https://doi.org/10.1073/pnas.022055499.

7   "Why Is the Atmospheric Carbon Reservoir So Small?," Earthguide, Geosciences Research Division, Scripps Institution of Oceanography, University of California San Diego, accessed March 3, 2020, http://earthguide.ucsd.edu/virtualmuseum/climatechange1/05_2.shtml.

8   "Daily CO_2," CO_2-earth, February 12, 2020, https://www.co2.earth/daily-co2.

9   Because of the greater land mass in the northern hemisphere, the worldwide CO_2 cycle is affected most by seasonal changes in the cycle there.

10 Jonathan Shieber, "CO_2 Levels in the Atmosphere Just Reached the Highest Level in Human History," World Economic Forum, May 16, 2019, https://www.weforum.org/agenda/2019/05/co2-in-the-atmosphere-just-exceeded-415-parts-per-million-for-the-first-time-in-human-history/.

11 United Nations Framework Convention on Climate Change, "Paris Agreement," December 12, 2015, https://unfccc.int/resource/docs/2015/cop21/eng/l09r01.pdf.

12 "EU Emissions Trading System (EU ETS)," European Commission, accessed March 3, 2020, https://ec.europa.eu/clima/policies/ets_en.

13 "A Brief Look at the Québec Cap-and-Trade-System for Emission Allowances," Government of Quebec, accessed March 3, 2020, http://www.mddep.gouv.qc.ca/changements/carbone/documents-spede/in-brief.pdf.

14 "Programmes découlant du Plan d'action 2013–2020 sur les changements climatiques," Conseil de gestion du Fonds vert, Government of Quebec, accessed March 3, 2020, http://www.environnement.gouv.qc.ca/cgfv/programmes.htm.

15 "Welcome," Regional Greenhouse Gas Initiative, accessed March 3, 2020, https://www.rggi.org/.

16 Testimony of Michael McSweeney, president and chief executive officer of the Cement Association of Canada, "Proceedings of the Standing Senate Committee on Energy, the Environment and Natural Resources," March 30, 2017, https://sencanada.ca/en/Content/Sen/Committee/421/ENEV/23ev-53196-e.

17 "British Columbia's Carbon Tax," Government of British Columbia, accessed March 3, 2020, https://www2.gov.bc.ca/gov/content/environment/climate-change/planning-and-action/carbon-tax.

18 Kim Willsher, "Macron Scraps Fuel Tax Rise in Face of Gilets Jaunes Protests," The Guardian, December 5, 2018, https://www.theguardian.com/world/2018/dec/05/france-wealth-tax-changes-gilets-jaunes-protests-president-macron.

19 Brad Plumer, "New U.N. Climate Report Says Put a High Price on Carbon," The New York Times, October 8, 2018, https://www.nytimes.com/2018/10/08/climate/carbon-tax-united-nations-report-nordhaus.html.

20 Richard Flanagan, "Australia Is Committing Climate Suicide," The New York Times, January 3, 2020, https://www.nytimes.com/2020/01/03/opinion/australia-fires-climate-change.html?action=click&module=Opinion&pgtype=Homepage.

21 World Bank Group, State and Trends of Carbon Pricing, 2019, accessed March 3, 2020, 41, http://documents.worldbank.org/curated/en/191801559846379845/pdf/State-and-Trends-of-Carbon-Pricing-2019.pdf.

22 "Portugal Produces Best Figures for Cutting EU Carbon Emissions," Business Live, May 8, 2019, https://www.businesslive.co.za/bd/world/europe/2019-05-08-portugal-produces-best-figures-for-cutting-eu-carbon-emissions/.

23   Mike Crawley, "What the Carbon Pricing Future Looks Like in Doug Ford's Ontario," CBC July 6, 2018, https://www.cbc.ca/news/canada/toronto/ontario-carbon-tax-doug-ford-justin-trudeau-1.4734971.

24   "Liberals Plan to Soften Carbon Tax Plan over Competitiveness Concerns," CBC News, August 1, 2018, https://www.cbc.ca/news/politics/liberals-carbon-price-lower-1.4769530.

25   For an excellent, accessible explanation of the Canadian situation, see Rachel Chen, "Everything You Need to Know about a Carbon Tax—and How It Would Work in Canada," *Chatelaine*, July 2, 2019, https://www.chatelaine.com/living/politics/federal-carbon-tax-canada/.

26   John Schwartz, "New Group, With Conservative Credentials, Plans Push for a Carbon Tax," *The New York Times*, June 19, 2018, https://www.nytimes.com/2018/06/19/climate/carbon-tax-climate-change.html; and Americans for Carbon Dividends, accessed March 3, 2020, https://www.afcd.org/.

27   Organisation for Economic Cooperation and Development (OECD), *Effective Carbon Rates 2018*, September 18, 2018, 15, https://read.oecd-ilibrary.org/taxation/effective-carbon-rates-2018_9789264305304-en#page17.

28   Plumer, "New U.N. Climate Report."

29   Nathaniel Rich, "Losing Earth," *The New York Times Magazine*, August 1, 2018, https://www.nytimes.com/interactive/2018/08/01/magazine/climate-change-losing-earth.html.

30   Zeke Hausfather, "Analysis: Global CO_2 Emissions Set to Rise 2% in 2017 after Three-Year 'Plateau,'" Carbon Brief, November 13, 2017, https://www.carbonbrief.org/analysis-global-co2-emissions-set-to-rise-2-percent-in-2017-following-three-year-plateau.

31   "Global Emissions of CO_2 from Fossil Fuels: 1900–2004," World Resources Institute, December 2005, http://www.wri.org/resources/charts-graphs/global-emissions-co2-fossil-fuels-1900-2004.

32   Glen Peters and Robbie Andrew, "Global CO_2 Emissions Likely to Rise in 2017," Center for International Climate Research, November 13, 2017, http://www.cicero.oslo.no/en/posts/klima/global-co2-emissions-likely-to-rise-in-2017.

33   Damian Carrington, "'Brutal News': Global Carbon Emissions Jump to All-Time High in 2018," *The Guardian*, December 5, 2018, https://www.theguardian.com/environment/2018/dec/05/brutal-news-global-carbon-emissions-jump-to-all-time-high-in-2018; Chelsea Harvey, "CO_2 Emissions Reached an All-Time High in 2018," *Scientific American*, December 6, 2018, https://www.scientificamerican.com/article/co2-emissions-reached-an-all-time-high-in-2018/.

34   Peters and Andrew, "Global CO_2 Emissions."

35   Robbie M. Andrew, "Global CO_2 Emissions from Cement Production," *Earth System Science Data* 10 (2018): 195–217, https://doi.org/10.5194/essd-10-195-2018; Madeleine Rubenstein, "Emissions from the Cement Industry," State of the Planet, May 9, 2012, http://blogs.ei.columbia.edu/2012/05/09/emis-

sions-from-the-cement-industry/; and "Global Emissions," Center for Climate and Energy Solutions, accessed March 3, 2020, https://www.c2es.org/content/international-emissions/.

36 The Covid-19 pandemic could turn out to be the catalyst for such a change. This book was being finalized for publication during the first months of the pandemic, when the long-term environmental and economic effects of the virus were unclear.

37 European Environment Agency, "Why Did Greenhouse Gas Emissions Decrease in the EU between 1990 and 2012?" Report of the European Environment Agency, 2014, https://www.eea.europa.eu/publications/why-are-greenhouse-gases-decreasing; Mathew L. Wald, "Carbon Dioxide Emissions Dropped in 1990, Ecologists Say," *The New York Times*, December 9, 1991, https://www.nytimes.com/1991/12/08/world/carbon-dioxide-emissions-dropped-in-1990-ecologists-say.html#story-continues-1; and Irina Kurganova, et al., "Carbon Cost of Collective Farming Collapse in Russia," *Global Change Biology* 20, no. 3 (2013), https://doi.org/10.1111/gcb.12379.

38 Andrew, "Global CO_2 Emissions from Cement Production," figures D1,D2,D3.

39 J.-F. Mercure, H. Pollitt, J.E. Viñuales, N.R. Edwards, P.B. Holden, U. Chewpreecha, P. Salas, I. Sognnaes, A. Lam, and F. Knobloch, "Macroeconomic impact of stranded fossil fuel assets," *Nature Climate Change* 8 (2018): 588–93, https://doi.org/10.1038/s41558-018-0182-1.

40 Boqiang Lin, Zihan Zhang, and Fei Ge, "Energy Conservation in China's Cement Industry," *Sustainability* 9 (2017): 668, doi:10.3390/su9040668.

41 Robert Wilson, "Building China: The Role of Cement in China's Rapid Development," Carbon Counter, June 6, 2015, originally posted on The Energy Collective, March 5, 2015, https://carboncounter.wordpress.com/2015/06/06/building-china-the-role-of-cement-in-chinas-rapid-development/.

42 Lin, Zhang, and Ge, "Energy Conservation in China's Cement Industry"; figure calculated from "Cement production globally and in the U.S. from 2010 to 2019 (in million metric tons)," Statista, accessed March 3, 2020, https://www.statista.com/statistics/219343/cement-production-worldwide/.

43 "Major countries in worldwide cement production from 2015 to 2019 (in million metric tons)," Statista, accessed March 3, 2020, https://www.statista.com/statistics/267364/world-cement-production-by-country/. China still has more people, but the population gap is decreasing—in 2019 China's population was an estimated 1.42 billion while India's was 1.35 billion. By way of comparison, the United States (population 321 million) produced 88.5 million tons of cement and Canada (population 35.85 million), about 12 million tons.

44 Michael Massing, "Does Democracy Avert Famine?," *The New York Times*, March 1, 2003, https://www.nytimes.com/2003/03/01/arts/does-democracy-avert-famine.html.

45 Amartya Sen, "Why India Trails China," *The New York Times*, June 19, 2013, https://www.nytimes.com/2013/06/20/opinion/why-india-trails-china.html.

46 Preetam Kaushik, "How to tackle India's affordable housing challenge," World Economic Forum, May 23, 2016, https://www.weforum.org/agenda/2016/05/how-to-tackle-india-s-affordable-housing-challenge.

47 Saurabh Kumar, "Budget 2019: PM Modi's Flagship Affordable Housing Scheme, Pradhan Mantri Awas Yojana (Urban), May Get a Boost," Financial Expert, June 30, 2019, https://www.financialexpress.com/budget/budget-2019-pm-modis-flagship-affordable-housing-scheme-pradhan-mantri-awas-yojana-urban-may-get-a-boost/1623493/. Amounts of housing, in the Indian numbering system, are 2.95 crore rural houses and 1.2 crore urban houses.

48 Pratyush Deep Kotoky, "Need to Speed Up Work on Urban Housing," *The Sunday Guardian*, August 17, 2019, https://www.sundayguardianlive.com/news/need-speed-work-urban-housing. Amount in Indian numbering system: Rs 32,877.63 crore.

49 2010 Brazilian census, quoted in "Favela Life: Rio's City Within a City," BBC News, June 9, 2014, https://www.bbc.com/news/world-latin-america-27635554#story_continues_2.

50 Gregory Scruggs, "Ministry of Cities RIP: The Sad Story of Brazil's Great Urban Experiment," *The Guardian*, July 18, 2019, https://www.theguardian.com/cities/2019/jul/18/ministry-of-cities-rip-the-sad-story-of-brazils-great-urban-experiment.

51 "20% Low-Cost Units Built under Urban Housing Schemes Vacant," *Times of India*, August 3, 2016, https://timesofindia.indiatimes.com/city/delhi/20-low-cost-units-built-under-urban-housing-schemes-vacant/articleshow/53525822.cms.

52 Saudamini Das, Arup Mitra, and Rajnish Kumar, "Do Neighborhood Facilities Matter for Slum Housing? Evidence from Indian Slum Clusters," *Urban Studies* 54, no. 8 (2016): 1887–1904, https://doi.org/10.1177/0042098016634578.

53 The Pritzker Architecture Prize, accessed March 3, 2020, https://www.pritzkerprize.com/laureates/balkrishna-doshi.

54 "Madhya Pradesh Poverty, Growth, and Inequality (English)," The World Bank, May 25, 2016, http://documents.worldbank.org/curated/en/454421467996665040/Madhya-Pradesh-poverty-growth-and-inequality.

55 Rahul Srivastava, "Arnaya: A Story of Incremental Growth," URBZ, August 8, 2011, http://www.urbz.net/articles/aranya-story-incremental-development.

56 "Aranya Population – Indore, Madhya Pradesh," Census 2011, accessed March 3, 2020, http://www.census2011.co.in/data/village/476343-aranya-madhya-pradesh.html.

57 Sukjong Hong, "Can Half a Good House Become a Home?," *The New Republic*, June 14, 2016, https://newrepublic.com/article/134223/can-half-good-house-become-home.

58 Montréal en statistiques, *Arrondissement du Plateau-Mont-Royal,* Ville de Montréal, May 2018, http://ville.montreal.qc.ca/pls/portal/docs/PAGE/MTL_STATS_FR/MEDIA/DOCUMENTS/PROFIL_SOCIOD%C-9MO_PLATEAU%20MONT-ROYAL%202016.PDF. It should be noted that the agglomeration of Montreal has a population density that is much less—3,779.8 per kilometre. Mile End is used as an example of how liveable neighbourhoods can be quite dense; even Outremont, where I live and which was originally built as a "garden suburb," has a density of 6,121 per kilometre. "Population totale en 2006 et en 2011—Variation—Densité," Ville de Montréal, accessed March 3, 2020, http://ville.montreal.qc.ca/pls/portal/docs/PAGE/MTL_STATS_FR/MEDIA/DOCUMENTS/01_POPULATION_DEN-SIT%C9_2011.PDF.

59 "Chandigarh Population 2011–2020 Census," Census 2011, accessed March 3, 2020, http://www.census2011.co.in/census/state/chandigarh.html.

60 "The Largest Cities in the World by Land Area, Population and Density," City Mayors Statistics, accessed March 3, 2020, http://www.citymayors.com/statistics/largest-cities-density-125.html.

61 Robbie Andrews, "Guest Post: Why India's CO_2 Emissions Grew Strongly in 2017," Carbon Brief, March 28, 2018, https://www.carbonbrief.org/guest-post-why-in-dias-co2-emissions-grew-strongly-in-2017. Brazilian statistics from "Brazil—CO_2 emissions per capita," Knoema, accessed March 3, 2020, https://knoema.com/atlas/Brazil/CO2-emissions-per-capita. As of 2016 (the most recent year for which complete statistics are available) India had the fifth highest CO_2 emissions in the world (2.4 gigatons), after China (10.2 gigatons), the United States (5.3 gigatons), Brazil (4.6 gigatons), and the 28 nations of the European Union (3.5 gigatons). India trailed them all when it came to CO_2 emissions per capita with 1.8 tons per person, compared to 7.2 tons in China, 6.9 tons in the EU, and 2.23 tons in Brazil. The champion with by far and away the greatest margin is the United States, however, with a whopping 16.5 tons per capita.

62 Jared Farmer, *Trees in Paradise: A California History* (New York: W.W. Norton, 2013).

63 "History of Smog," *L.A. Weekly,* September 22, 2005, http://www.laweekly.com/news/history-of-smog-2140714; and "About," South Coast Air Quality Management District, accessed March 3, 2020, http://www.aqmd.gov/nav/about.

64 José Ernesto Credendio and Mario Cesar Carvalho, "Rodoanel não vai aplacar tráfego lento em SP, aponta estudo," *Folha de S.Paulo,* March 26, 2014; and "Rodoanel Mário Covas – problemas com 'ovo e a galinha,'" Blog da Cargobr, March 26, 2014, http://blog.cargobr.com/rodoanel-problemas-2/: the blog of Cargobr, a major freight carrier.

65 Georgina Santosa, Hannah Behrendt, Laura Maconic, Tara Shirvanic, and Alexander Teytelboym, "Part I: Externalities and economic policies in road

transport," *Research in Transportation Economics* 28, no. 1 (2010): 2–45, https://doi.org/10.1016/j.retrec.2009.11.002.

66   Calculated by Aviation Environment Federation, a UK-based organization campaigning for aviation's impacts on people and the environment to be brought within sustainable limits. 2008 figures: "Kilograms of CO_2 per passenger kilometre for different modes of transport within the UK," Aviation Environment Federation, accessed March 3, 2020, http://www.aef.org.uk/downloads/Grams_CO2_transportmodesUK.pdf.

67   Amritha Pillay, "Petcoke Imports on a Rise as Coal Substitute Largely Due to Users Like Cement Companies Opting for Petcoke as Cheaper Option Instead of Coal," *Business Standard*, May 23, 2017, https://www.business-standard.com/article/economy-policy/petcoke-imports-on-a-rise-as-coal-substitute-117052200624_1.html; and Harish V. Nair, "Supreme Court to Centre: Pet Coke, Furnace Oil Main Causes of Pollution in Delhi-NCR, Must Be Banned," *India Today*, February 7, 2017, https://www.indiatoday.in/mail-today/story/air-pollution-delhi-pet-coke-furnance-oil-959344-2017-02-07.

68   Aditi Roy Ghatak and Karl Mathiesen, "US Exports of Tar Sands Waste Are Fuelling Delhi's Air Pollution Crisis," Climate Change News, September 28, 2017, http://www.climatechangenews.com/2017/09/28/us-exports-tar-sand-waste-fuelling-delhis-air-pollution-crisis/; "India Bans Petroleum Coke Import for Use as Fuel," *The Economic Times*, August 17, 2018, https://economictimes.indiatimes.com/industry/indl-goods/svs/metals-mining/india-bans-petroleum-coke-import-for-use-as-fuel/articleshow/65439259.cms.

69   OECD definitions: "Petroleum Coke," OECD Glossary of Statistical Terms, March 11, 2003, https://stats.oecd.org/glossary/detail.asp?ID=4618.

70   Yuli Shan, Dabo Guan, Jing Meng, Zhu Liu, Heike Schroeder, Jianghua Liu, and Zhifu Mi, "Rapid Growth of Petroleum Coke Consumption and Its Related Emissions in China," *Applied Energy* 226, no. 15 (2018): 494–502, https://doi.org/10.1016/j.apenergy.2018.06.019.

71   Oliver, et al., "Carbon, Fossil Fuel, and Biodiversity Mitigation."

72   "Calera: Proven, Economical, and Better for the Planet," Calera, accessed March 5, 2020, http://www.calera.com/beneficial-reuse-of-co2/science.html.

73   "Technology," Blue Planet, accessed March 5, 2020, http://www.blueplanet-ltd.com/#technology.

74   "How the oil sands and XPRIZE could reinvent carbon," XPRIZE, accessed March 5, 2020, https://carbon.xprize.org/press-release/ten-teams-five-countries-advance-finals-of-20m-nrg-cosia-carbon-xprize?language=en.

75   Chris Stern, email message to author, July 23, 2019.

76   Kate Baggaley, "'Green' Concrete Could Be Game-Changer for Construction Industry Microscopic Flakes of Graphene Add Strength and Durability—But Also Raise Cost and Safety Concerns," NBC News, May 2, 2018, https://www.

nbcnews.com/mach/science/new-green-concrete-could-be-game-changer-construction-industry-ncna870371.

77　Low Emissions Intensity Lime and Cement, accessed March 3, 2020, www.leilac.org.uk.

78　We Don't Have Time, "Concrete Ways to Solve Our Concrete Problem," Medium, April 9, 2019, https://medium.com/wedonthavetime/concrete-ways-to-solve-our-concrete-problem-ae307d64d2fe; and Aaron McArthur, "Vancouver Developer Proposes World's Tallest Wood Tower," Global News, April 25, 2019, https://globalnews.ca/news/5201857/vancouver-tallest-wood-tower/.

79　mgb ARCHITECTURE + DESIGN, Equilibrium Consulting, LMDG Ltd., and BTY Group, "The Case for Tall Wood Buildings," 26.

80　Gabriel Popkin, "How Much Can Forests Fight Climate Change? Trees Are Supposed to Slow Global Warming, But Growing Evidence Suggests They Might Not Always Be Climate Saviours," *Nature*, January 15, 2019, https://www.nature.com/articles/d41586-019-00122-z.

81　Bill McKibben, "Don't Burn Trees to Fight Climate Change—Let Them Grow," *The New Yorker*, August 15, 2019, https://www.newyorker.com/news/daily-comment/dont-burn-trees-to-fight-climate-changelet-them-grow.

82　Sarah Zhang, "Europe Is Running Low on CO_2: The Shortfall Has Sparked Fears of Shortages of Beer, Meat, and Crumpets. (Crumpets!)," *The Atlantic*, July 6, 2018, https://www.theatlantic.com/science/archive/2018/07/europes-co2-shortage/564458/.

83　"China Shows Climate Change Prowess as Large-Scale CCUS Facility Enters Construction," Markets Insider, March 29, 2017, https://markets.businessinsider.com/news/stocks/china-shows-climate-change-prowess-as-large-scale-ccus-facility-enters-construction-1001879901.

84　Sonal Patel, "Capturing Carbon and Seizing Innovation: Petra Nova Is POWER's Plant of the Year," POWER, August 1, 2017, http://www.powermag.com/capturing-carbon-and-seizing-innovation-petra-nova-is-powers-plant-of-the-year/?pagenum=1.

85　"Top 10 Cement Producer Profiles," *Global Cement Magazine*, July–August 2018, 7, https://globalcement.com/magazine/articles/1072-top-10-cement-producer-profiles. The number two producer, at 8 percent, is LafargeHolcim, which operates around the world from its European base. Number three is the Chinese firm Anhhui Conch (7.5 percent), which sells primarily in China, Southeast Asia, and Russia.

86　Natural Resources Defense Council "The Road from Paris: China's Progress toward Its Climate Pledge," Issue brief 17-11-D, November 2017, https://www.nrdc.org/sites/default/files/paris-climate-conference-China-IB.pdf.

87　Jonathan Rowland, "China Drags on Global Cement Consumption Growth," *World Cement*, October 17, 2017, https://www.worldcement.com/special-reports/17102017/china-drags-on-global-cement-consumption-growth/.

88   Brad Plumer, Nadja Popovich, and Shola Lawal, "The Coronavirus and Carbon Emissions," *The New York Times*, February 26, 2020, https://www.nytimes.com/2020/02/26/climate/nyt-climate-newsletter-coronavirus.html.

89   Charles C. Mann, "Renewables Aren't Enough: Clean Coal Is the Future," *Wired*, March 25, 2014, https://www.wired.com/2014/03/clean-coal/.

90   "Instructions for Performing a Multifamily Property Condition Assessment (Version 2.0)," Appendix F, Fannie Mae, 2014, https://www.fannimae.com/content/guide_form/4099f.pdf.

91   Jennifer O'Connor, "Survey on Actual Service Lives for North American Buildings," (paper presented at Wood Frame Housing Durability and Disaster Issues Conference, Las Vegas, NV, October, 2004), http://cwc.ca/wp-content/uploads/2013/12/DurabilityService_Life_E.pdf.

92   Daniel Tencer, "Toronto Skyline Evolution: Video Shows How City Is Transforming Amid Skyscraper Boom," Huffington Post, April 4, 2019, https://www.huffingtonpost.ca/2019/04/04/toronto-skyline-evolution_a_23706558/. Among these skyscrapers is Monde, a forty-four-storey luxury condo development designed in part by Habitat 67 architect Moshe Safdie. It features many units with gardens but is far from "affordable housing." One-bedroom units are listed beginning at $574,990. "Monde Residential Development," Safdie Architects, accessed March 3, 2020, https://www.safdiearchitects.com/projects/monde-residential-development.

93   See ten of them: "Top 10 U.S. Stadium/Arena Demolitions," YouTube video, 5:37, posted by Richard S. Dargan, March 15, 2015, https://www.youtube.com/watch?v=7X7hWcctXLw.

94   Among the problems: a concrete chunk weighing fifty-five tons fell off in 1991, making the Montreal Expos, who used the "Big O" as their home field, finish the season playing all their games away. Seven years later heavy snow caused the roof to tear; that time the Rolling Stones had to cancel two concerts. See "Roger Taillibert défend la conception du Stade Olympique," Radio-Canada, July 12, 2016, https://ici.radio-canada.ca/nouvelle/792320/stade-olympique-architecte.

95   Zaria Gorvett, "Will the Skyscrapers Out Last the Pyramids?," BBC, August 9, 2016, http://www.bbc.com/future/story/20160808-will-the-skyscrapers-out-last-the-pyramids.

96   Tansy Hoskins, "Reliving the Rana Plaza Factory Collapse," *The Guardian*, April 23, 2015, https://www.theguardian.com/cities/2015/apr/23/rana-plaza-factory-collapse-history-cities-50-buildings.

97   Forty, *Concrete and Culture*, 76.

98   Courtney Humphries, "Why We Should Let the Pantheon Crack: Modern Architects Have a Lot to Learn from the Sound Engineering of the Ancients," *Nautilus*, May 21, 2015, nautil.us/issue/24/error/why-we-should-let-the-pantheon-crack.

99   Forty, *Concrete and Culture*, 52.

100 "Concrete Conservation," The Getty Conservation Institute, November 2017, http://www.getty.edu/conservation/our_projects/field_projects/concrete/overview.html.

101 Ibid.

102 Elif Batuman, "The Sanctuary: The World's Oldest Temple and the Dawn of Civilization," *The New Yorker*, December 19 and 26, 2011, https://www.newyorker.com/magazine/2011/12/19/the-sanctuary.

103 Forty, *Concrete and Culture*, 80.

104 Paul Jannot, quoted in Forty, *Concrete and Culture*, 32.

105 Jenni Marsh, "Tadao Ando: The Japanese Boxer Turned Pritzker Prize Winner Who Buried the Buddha," CNN, November 5, 2017, https://www.cnn.com/style/article/tadao-ando-exhibition/index.html; and Martin Laflamme, "Tadao Ando: When Every Single Building Is a Passion Project," *Japan Times*, October 29, 2017, https://www.japantimes.co.jp/life/2017/10/29/style/tadao-ando-every-single-building-passion-project/#.WyoH6KknZ8w.

106 Wendy Lesser, *You Say to Brick: The Life of Louis Kahn* (New York: Farrar, Straus and Giroux, 2017), 225.

107 Ibid., 192.

108 Forty, *Concrete and Culture*, 134.

109 "MFE Formwork Transforms 'Impossible' Building Projects into Modern Construction Marvels," *South China Morning Post*, December 27, 2017, http://www.scmp.com/country-reports/business/topics/malaysia-business-report-2017/article/2125191/mfe-formwork.

110 This quote is the origin of the title of Wendy Lesser's wonderful biography: *You Say to Brick*. Note too that Kahn sometimes mumbled, so much so that in a film about him by his son Nathaniel (*My Architect: A Son's Journey*) the Spanish subtitles translate him as if he said "oranges" not "arches." Try to make sense of that! *My Architect: A Son's Journey*, directed by Nathaniel Kahn (Louis Kahn Project Inc., Mediaworks, 2004).

111 *Art & Architecture Thesaurus Online*, s.v. "International Style," accessed March 3, 2020, http://www.getty.edu/vow/AATFullDisplay?find=international+-style&logic=AND&note=&page=1&subjectid=300021472.

112 Anna Winston, "Moshe Safdie used all the Lego in Montreal to design Habitat 67," *Dezeen*, December 19, 2014, https://www.dezeen.com/2014/12/19/moshe-safdie-movie-interview-habitat-67/.

113 Alison Furuto, "National Museum of Afghanistan Competition Winners," *Architecture Daily*, October 3, 2012, https://www.archdaily.com/278348/national-museum-of-afghanistan-competition-winners.

114 Rory Stott, "Zaha Hadid Architects Designs Parabolic-Vaulted School Campus in Rural China," The Concrete Centre, April 19, 2018, https://www.concrete-centre.com/Publications-Software/Concrete-Quarterly-New/News.aspx.

115 "AD Interviews: Moshe Safdie."

116 Natasha Ann Zachariah, "Quality of Life Has a Price, Says Moshe Safdie," *The Straits Times*, August 1, 2015, https://www.straitstimes.com/lifestyle/home-design/quality-of-life-has-a-price-says-moshe-safdie.

117 "AD Interviews: Moshe Safdie."

118 Ruth La Ferla, "A Day Out with the Architect Moshe Safdie," *The New York Times*, December 30, 2015, https://www.nytimes.com/2015/12/31/style/a-day-out-with-the-architect-moshe-safdie.html.

119 Ibid.

120 Amy Plitter, "Bjarke Ingels Tapped for Replacement for Nomad's Erstwhile Bancroft Building: Moshe Safdie Was Originally Attached to Design the Building at 3 West 29th Street," *Curbed New York*, September 29, 2017, https://ny.curbed.com/2017/9/29/16383972/bjarke-ingels-group-nyc-building-hfz-capital.html; Nikolai Fedak, "Bjarke Ingels-Designed '29th & 5th' Revealed, HFZ Capital's New NoMad Office Tower at 3 West 29th Street," New York Yimby, April 3, 2018, https://newyorkyimby.com/2018/04/bjarke-ingels-designed-29th-5th-revealed-hfz-capitals-new-nomad-office-tower-at-3-west-29th-street.html; and Rich Bockmann, "Feldman's HFZ adds to NoMad assemblage with $30M buy," The Real Deal, July 26, 2018, https://therealdeal.com/2018/07/26/hfz-adds-to-nomad-assemblage-with-30m-buy/?_ct=30ne93im7mbj.

121 Drawn from the excellent book by Laurence Cossé, *La grande arche*. It is described in French as a *roman*, a novel, but it really is what in English is called creative non-fiction. Carefully researched, it also contains Cossé's reading of the tragedy—or triumph—of the arch. Laurence Cossé, *La grande arche* (Paris: Gallimard, 2016).

122 Brigit de Kosmi, quoted in Cossé, *La grande arche*, 210. My translation.

123 Robert Lion, quoted in Cossé, *La grande arche*, 210. My translation.

124 Ibid., 292.

125 Moshe Safdie, personal communication, August 6, 2018.

126 "AD Interviews: Moshe Safdie."

CHAPTER 6

1 A few weeks later a much-touted US$200 billion plan was unveiled, but at this writing was still stalled: in September 2019 Trump refused to move forward on a plan with bipartisan support because Democrats were pushing for more investigation of his activities. See John S. Tobey, "Infrastructure: The Forgotten Plan That Could Prevent Recession, Spark Stocks and Reelect Trump," *Forbes*, September 13, 2019, https://www.forbes.com/sites/johntobey/2019/09/13/infrastructure-the-forgotten-plan-that-could-prevent-recession-spark-stocks-and-reelect-trump/#4491484b5cb6; see also U.S. Secretary of Transport Elaine Chao explaining the snags in July 2018 in a *Fortune Magazine* forum: http://fortune.com/video/2018/07/17/what-happened-to-trumps-infrastructure-plans/.

2   Aaron Kirchfeld and Scott Deveau, "Caisse Hires Advisers to Explore McInnis Cement Sale," *Bloomberg*, January 4, 2018, https://www.bloomberg.com/news/articles/2018-01-04/caisse-is-said-to-hire-advisers-to-explore-mcinnis-cement-sale.

3   "McInnis Cement Closes a Refinancing to Support Its Growth and Expansion," July 17, 2019, https://mcinniscement.com/press-release/mcinnis-cement-closes-refinancing/.

4   Stéphanie Gendron, "Gaspésie: une partie de la 132 part dans le fleuve: Les grandes marées et les forts vents ont causé beaucoup de dommages," *Journal de Québec*, December 16, 2016, http://www.journaldequebec.com/2016/12/16/gaspesie-les-dommages-causes-par-les-grandes-marees-forcent-la-fermeture-de-la-route-132-sur-une-distance-de-60-km; and https://ici.radio-canada.ca/nouvelle/1082545/erosion-des-berges-les-changements-climatiques-ne-sont-pas-les-seuls-responsables.

5   "Discover Lachine," Ville de Montréal, accessed March 5, 2020, http://ville.montreal.qc.ca/portal/page?_pageid=8197,90911629&_dad=portal&_schema=PORTAL.

6   Isabelle Porter, "Les exilés de l'érosion des berges du Saint-Laurent," *Le Devoir*, June 29, 2019, https://www.ledevoir.com/societe/environnement/557735/les-exiles-de-l-erosio-demenager-ou-attendre-la-tempete.

SELECTED BIBLIOGRAPHY

REPORTS

Centre d'information sur le ciment et ses applications. "Histoire du béton: Naissance et développement, de 1818 à nos jours." Cahier des modules de conference pour les écoles d'architecture, B90A. 2009. http://www.infociments.fr/telecharger/CT-B90A.pdf

European Environment Agency. "Why Did Greenhouse Gas Emissions Decrease in the EU between 1990 and 2012?" Report of the European Environment Agency, 2014. https://www.eea.europa.eu/publications/why-are-greenhouse-gases-decreasing.

Genivar Inc. "Projet d'implantation d'une cimenterie sur le territoire de Port-Daniel-Gascons: Mise à jour de l'étude de répercussion sur l'environnement. Document consolidé pour diffusion publique intégrant le rapport original et les analyses supplémentaires demandées par le MDDEFP." Montreal: GENIVAR Inc., 2013.

mgb ARCHITECTURE + DESIGN, Equilibrium Consulting, LMDG Ltd., and BTY Group. "The Case for Tall Wood Buildings: How Mass Timber Offers a Safe, Economical, and Environmentally Friendly Alternative for Tall Building Structures." 2012. http://cwc.ca/wp-content/uploads/publications-Tall-Wood.pdf.

Natural Resources Defense Council. "The Road from Paris: China's Progress toward Its Climate Pledge." Issue brief 17-11-D,

November 2017. https://www.nrdc.org/sites/default/files/paris-climate-conference-China-IB.pdf.

Park and Recreation Department, City of San Diego, and Merkel & Associates, Inc. "San Diego River Natural Resource Management Plan." Final draft, May 2006. https://nrm.dfg.ca.gov/FileHandler.ashx?DocumentID=24784.

United Nations Environment Programme. "Sand, Rarer Than One Thinks." UNEP Global Environment Alert Service report, March 2014. GEAS_Mar2014_Sand_Mining.pdf.

United Way Toronto. "Poverty by Postal Code 2: Vertical Poverty—Declining Income, Housing Quality and Community Life in Toronto's Inner Suburban High-Rise Apartments." 2001. https://www.unitedwaygt.org/document.doc?id=89.

BOOKS

Alexievich, Svetlana. *Voices from Chernobyl: The Oral History of a Nuclear Disaster*. Translated by Keith Gessen. London: Picador, 2006.

Benoit, Michèle, and Roger Gratton. *Pignon sur rue: Les quartiers de Montréal*. Montreal: Guerin éditeur, 1991.

Collins, Peter. *Concrete: The Vision of a New Architecture*. Montreal and Kingston: McGill-Queen's University Press, 2004.

Cossé, Laurence. *La grande arche*. Paris: Gallimard, 2016.

Courland, Robert. *Concrete Planet: The Strange and Fascinating Story of the World's Most Common Man-Made Material*. Amherst, NY: Prometheus, 2011.

Didion, Joan. *The White Album*. New York: Simon and Schuster, 1979.

Donnell, Clayton. *The Fortifications of Verdun, 1874–1917*. Oxford: Osprey Editions, 2011.

Farmer, Jared. *Trees in Paradise: A California History*. New York: W W Norton, 2013.

Faulkner, William. *As I Lay Dying*. New York: The Modern Library, 1930.

Florman, Samuel C. *The Civilized Engineer*. New York: St. Martin's Press, 1987.

Forty, Adrian. *Concrete and Culture: A Material History.* London: Reaktion Books, 2012.

Fynn, Shaun, and Maristella Casciato. *Chandigarh Revealed: Le Corbusier's City Today.* New York: Princeton Architectural Press, 2017.

Gargiani. Roberto, *Concrete from Archeology to Invention, 1700–1769: The Renaissance of Pozzolana and Roman Construction Techniques.* Translated by Stephen Piccolo. Laussane: EOFL Press, 2013.

Gladkov, F.V. *Cement: A Novel.* Translated by A.S. Athur and C. Ashleigh. Evanston, IL: Northwestern University Press, 1994.

Grindrod, John. *Concretopia: A Journey Around the Rebuilding of Postwar Britain.* Brecon: Old Street Publishing, 2014.

Jacobs, Jane. *The Death and Life of Great American Cities.* New York: Vintage, 1992.

Kosmatka, Steven H., and Michelle L. Wilson. *Design and Control of Concrete Mixtures.* 16th ed. Skokie, IL: Portland Cement Association, 2016.

Kuhn, Gerald G., and Francis P. Shepard. *Sea Cliffs, Beaches, and Coastal Valleys of San Diego County: Some Amazing Histories and Some Horrifying Implications.* Berkeley and Los Angeles: University of California Press, 1984.

Lancaster, Lynne C. *Concrete Vaulted Construction in Imperial Rome: Innovations in Context.* Cambridge: Cambridge University Press, 2005.

Langer, Lawrence L. *Pre-empting the Haulocaust.* New Haven, CT: Yale University Press, 1998.

Le Corbusier. *The City of To-morrow and Its Planning.* Translated by Fredrerick Etchells. New York: Dover Publications, 1987. First published 1929 by Payson & Clarke, New York.

Lee Kuan Yew, *From Third World to First: The Singapore Story: 1965–2000.* New York: HarperCollins, 2000.

Leopold, Aldo. *A Sand County Almanac.* New York: Ballantine Books, 1970.

Lesser, Wendy. *You Say to Brick: The Life of Louis Kahn.* New York: Farrar, Straus and Giroux, 2017.

Mukerji, Chandra. *Impossible Engineering: Technology and Territoriality on the Canal du Midi.* Princeton, NJ: Princeton University Press, 2009.

Murphy, Edward Charles, and others. *Destructive Floods in the United States in 1905 with a Discussion of Flood Discharge and Frequency and an Index to Flood Literature.* Washington, DC: Government Printing Office, 1906. https://pubs.usgs.gov/wsp/0162/report. pdf.

Owen, David. *Where the Water Goes: Life and Death along the Colorado River.* New York: Riverhead Books, 2017.

Perry, Martin, Lily Kong, and Brenda Yeoh, *Singapore: A Developmental City State.* New York: John Wiley and Sons, 1997.

Pitzer, Paul C. *Grand Coulee: Harnessing a Dream.* Pullman, WA: Washington State University Press, 1994.

Possehl, Gregory L. *The Indus Civilization: A Contemporary Perspective.* New York: AltaMira Press, 2002.

Scott, Robert A. *The Gothic Enterprise: A Guide to Understanding the Medieval Cathedral.* Berkeley: University of California Press, 2003.

Shepard, Wade. *Ghost Cities of China: The Story of Cities without People in the World's Most Populated Country.* London: Zed Books, 2015.

Soderstrom, Mary. *Green City: People, Nature and Urban Places.* Montreal: Véhicule Press, 2007

———. *The Walkable City: From Haussmann's Boulevards to Jane Jacobs' Streets and Beyond.* Montreal: Véhicule Press, 2008.

———. *Road Through Time: The Story of Humanity on the Move.* Regina: University of Regina Press, 2017.

Stegner, Wallace. *Angle of Repose.* New York: Doubleday, 1971.

Steinbeck, John. *The Grapes of Wrath.* New York: Penguin, 2006.

Tapper, Richard, and Keith McLachlan, eds. *Technology, Tradition and Survival: Aspects of Material Culture in the Middle East and Central Asia.* Abingdon: Routledge, 2002.

Thompson, Erwin N. *The Guns of San Diego: San Diego Harbor Defenses, 1796–1947.* Edited by Howard B. Overton. San Diego:

National Park Service, 1991. https://www.nps.gov/parkhistory/online_books/cabr/index.htm.

Wallace-Handrill, Andrew. *Houses and Society in Pompeii and Herculaneum*. Princeton, NJ: Princeton University Press, 1994.

Weber, Nicholas Fox. *Le Corbusier: A Life*. New York: Alfred A. Knopf, 2008.

Vicat, Louis. *Recherches expérimentales sur le chaux de construction, les bétons et les mortiers ordinaires*. Paris, 1817. https://archive.org/details/recherchesexperioovica.

Vitruvius, Marcus Pollio. *Ten Books on Architecture*. Translated by Joseph Gwilt. London: Longman, Brown, Green, and Longmans, 1826. https://archive.org/stream/architecturemaroogwilgoog/architecturemaroogwilgoog_djvu.txt

ARTICLES

Ali, Muhammad Shafqat, Muhammad Sagheer Aslam, and M. Saeed Mirza. "A Sustainability Assessment Framework for Bridges—A Case Study: Victoria and Champlain Bridges, Montreal." *Structure and InfraStructure Engineering* 12, no. 11 (2016): 1381–94. http://dx.doi.org/10.1080/15732479.2015.1120754.

Atlanta Preservation and Planning Services. "Introduction to Tabby." Compiled from Lauren B. Sickels-Taves and Michael S. Sheehan, *The Lost Art of Tabby: Preserving Oglethorpe's Architectural Legacy*, Southfield, MI: Architectural Conservation Press, 1999. Atlanta Preservation. http://atlantapreservation.com/buildingmaterials/TabbyInfo.pdf.

Batuman, Elif. "The Sanctuary: The World's Oldest Temple and the Dawn of Civilization." *The New Yorker*, December 19 and 26, 2011, https://www.newyorker.com/magazine/2011/12/19/the-sanctuary.

Chen, Rachel. "Everything You Need to Know about a Carbon Tax—And How It Would Work in Canada." *Chatelaine*, July 2, 2019. https://www.chatelaine.com/living/politics/federal-carbon-tax-canada/.

Cornwall, Warren. "Is Wood a Green Source of Energy? Scientists Are Divided." *Science*, January 5, 2017. doi:10.1126/science.aal0574.

Flanagan, Richard. "Australia Is Committing Climate Suicide." *The New York Times*, January 3, 2020. https://www.nytimes.com/2020/01/03/opinion/australia-fires-climate-change.html?action=click&module=Opinion&pgtype=Homepage.

Gendron, Stéphanie. "Gaspésie: une partie de la 132 part dans le fleuve: Les grandes marées et les forts vents ont causé beaucoup de dommages." *Journal de Québec*, December 16, 2016. http://www.journaldequebec.com/2016/12/16/gaspesie-les-dommages-causes-par-les-grandes-marees-forcent-la-fermeture-de-la-route-132-sur-une-distance-de-60-km.

Hohlfelder, R.L., C. Brandon, and J.P. Oleson. "Constructing the Harbour of Cesarea on the Sea: New Evidence from the ROMACONS Field Campaign of October 2005." *The International Journal of Nautical Archaeology* 36, no. 2 (2007): 409–15. https://web.uvic.ca/~jpoleson/ROMACONS/Caesarea2005.htm.

Hoshowsky, Robert. "The Future of Cement in North America: McInnis." *Business in Focus Magazine*, August 2016. https://www.businessinfocusmagazine.com/2016/08/the-future-of-cement-in-north-america/.

Hughes, J. Donald, and J.V. Thirgood. "Deforestation, Erosion, and Forest Management in Ancient Greece and Rome." *Journal of Forest History* 26, no. 2 (1982): 60–75. http://www.jstor.org/stable/4004530.

Kruse, Kevin M. "What Does a Traffic Jam in Atlanta Have to Do with Segregation? Quite a Lot." *The New York Times*, August 14, 2019. https://www.nytimes.com/interactive/2019/08/14/magazine/traffic-atlanta-segregation.html.

Kurganova, Irina, Valentin Lopes de Gerenyu, Johan Six, and Yakov Kuzyakov. "Carbon Cost of Collective Farming Collapse in Russia." *Global Change Biology* 20, no. 3 (2014): 938–47. https://doi.org/10.1111/gcb.12379.

Li, Rongchao. "Flood Control in the Yellow River Basin in China." *Water Encyclopedia*, April 15, 2005. https://doi.org/10.1002/047147844X.sw259.

Liu Xianzhao and Qi Shanzhong. "Wetlands Environmental Degradation in the Yellow River Delta, Shandong Province of China." *Procedia Environmental Sciences* 11 (2011): 701–5 doi:10.1016/j.proenv.2011.12.109.

La Ferla, Ruth. "A Day Out with the Architect Moshe Safdie." *The New York Times*, December 30, 2015. https://www.nytimes.com/2015/12/31/style/a-day-out-with-the-architect-moshe-safdie.html.

Madrigal, Alex. "The Racist Housing Policy that Made Your Neighborhood." *The Atlantic*, May 22, 2014. https://www.theatlantic.com/business/archive/2014/05/the-racist-housing-policy-that-made-your-neighborhood/371439/.

Marsh, Jenni. "Tadao Ando: The Japanese Boxer Turned Pritzker Prize Winner Who Buried the Buddha." CNN, November 5, 2017. https://www.cnn.com/style/article/tadao-ando-exhibition/index.html.

Micklin, Philip. "The Aral Sea Disaster." *Annual Review of Earth and Planetary Sciences* 35, no.1 (2007): 4772. doi:10.1146/annurev.earth.35.031306.140120.

Oliver, Chadwick Dearing, Nedal T. Nassar, Bruce R. Lippke, and James B. McCarter. "Carbon, Fossil Fuel, and Biodiversity Mitigation with Wood and Forests." *Forests, Journal of Sustainable Forestry* 33, no. 3 (2014): 248–75. doi:10.1080/10549811.2013.839386.

Penn, Ivan. "The $3 Billion Plan to Turn Hoover Dam into a Giant Battery." *The New York Times*, July 24, 2018. https://www.nytimes.com/interactive/2018/07/24/business/energy-environment/hoover-dam-renewable-energy.html.

Rich, Nathaniel. "Losing Earth." *The New York Times Magazine*, August 1, 2018. https://www.nytimes.com/interactive/2018/08/01/magazine/climate-change-losing-earth.html.

Safdie, Moshe. "Habitat Original Proposal." Item finding aid description. The Moshe Safdie Archive. McGill University Library. http://cac.mcgill.ca/moshesafdie/fullrecord.php?ID=11059&d=1.

Scruggs, Gregory. "Ministry of Cities RIP: The Sad Story of Brazil's Great Urban Experiment." The Guardian, July 18, 2019. https://www.theguardian.com/cities/2019/jul/18/ministry-of-cities-rip-the-sad-story-of-brazils-great-urban-experiment.

Sen, Amartya. "Why India Trails China." The New York Times, June 19, 2013. https://www.nytimes.com/2013/06/20/opinion/why-india-trails-china.html.

Stroberg, Joseph. "What Happens to All the Salt We Dump on the Roads." Smithsonian Magazine, January 6, 2014. https://www.smithsonianmag.com/science-nature/what-happens-to-all-the-salt-we-dump-on-the-roads-180948079/.

Wald, Mathew L. "Carbon Dioxide Emissions Dropped in 1990, Ecologists Say." The New York Times, December 9. 1991. https://www.nytimes.com/1991/12/08/world/carbon-dioxide-emissions-dropped-in-1990-ecologists-say.html#story-continues-1.

Walker, Thomas, Peter Cardellichio, John S. Gunn, David S. Saah, and John M. Hagan. "Carbon Accounting for Woody Biomass from Massachusetts (USA) Managed Forests: A Framework for Determining the Temporal Impacts of Wood Biomass Energy on Atmospheric Greenhouse Gas Levels." Journal of Sustainable Forestry 32, no. 1–2 (2013): 130–58. https://doi.org/10.1080/10549811.2011.652019.

Zachariah, Natasha Ann. "Quality of Life Has a Price, Says Moshe Safdie." The Straits Times, August 1, 2015. https://www.straitstimes.com/lifestyle/home-design/quality-of-life-has-a-price-says-moshe-safdie.

INDEX

PHOTO: ANNE RICHARD

ABOUT THE AUTHOR

Concrete: From Ancient Origins to a Problematic Future is MARY SODERSTROM's seventeenth book. She has published novels, short story collections, and non-fiction books, including her acclaimed *Road Through Time: The Story of Humanity on the Move* and most recently *Frenemy Nations: Love and Hate between Neighbo(u)ring States.*

ABOUT THE TYPE

This book is set in *Adobe Jenson,* an old-style serif typeface drawn for Adobe Systems by its chief type designer Robert Slimbach. Its Roman styles are based on a text face cut by Nicolas Jenson in Venice around 1470, and its italics are based on those created by Ludovico Vicentino degli Arrighi fifty years later.

The accents are set in *Knockout.* This family of sans serif typefaces expands upon Jonathan Hoefler's first typeface, Champion Gothic (1990). Like Champion, Knockout is inspired by a style of American wood type which was first introduced in the mid-nineteenth century.

www.ingramcontent.com/pod-product-compliance
Lightning Source LLC
Chambersburg PA
CBHW032120020426
42334CB00016B/1017